Contributors

Marion Bonhomme National Institute on Applied Sciences (INSA), Laboratoire Matériaux et Durabilité des Constructions (LMDC), Toulouse III University Toulouse, Toulouse, France

Rafael Calderón-Contreras Department of Social Sciences, Universidad Autónoma Metropolitana-Cuajimalpa, Mexico City, Mexico

Bikramaditya K. Choudhary CSRD/SSS, Jawaharlal Nehru University, New Delhi, India

Marco Córdova FLACSO Ecuador, Quito, Ecuador

Marcos Tavares de Arruda Filho Instituto de Energia E Ambiente—Universidade de São Paulo, São Paulo, Brasil

Brais Estévez Departamento de Xeografía, Universidade de Santiago de Compostela, Santiago de Compostela, Spain

Manuela Á. Fernández-Borrero COIDESO-Universidad de Huelva, Huelva, Spain

Mamdouh Ibrahim Zaky Mohamed Galil College of Engineering and Architecture, University of Nizwa, Nizwa, Oman

Bruno Gandlgruber Department of Institutional Studies, Universidad Autónoma Metropolitana-Cuajimalpa, Mexico City, Mexico

Farzaneh Gharaati Department of Architecture, Tarbiat Modares University, Tehran, Iran

Salomón González-Arellano Department of Social Sciences, Universidad Autónoma Metropolitana-Cuajimalpa, Mexico City, Mexico

José Luis González Madrigal Universidad Autonoma Metropolitana, Mexico City, Mexico

Julia Hidalgo National Centre for Scientific Research (CNRS), Laboratoire Interdisciplinaire Solidarités, Societés, Territoires (LISST), Toulouse II University, Toulouse, France

Marco Jaso Department of Institutional Studies, Universidad Autónoma Metropolitana-Cuajimalpa, Mexico City, Mexico

Daniela Losiggio CONICET, Universidad Nacional Arturo Jauretche, Buenos Aires, Argentina

Mohammadjavad Mahdavinejad College of Engineering and Architecture, University of Nizwa, Nizwa, Oman;
Department of Architecture, Tarbiat Modares University, Tehran, Iran

Blanca Miedes-Ugarte COIDESO-Universidad de Huelva, Huelva, Spain

Nora A. Morales Zaragoza Department of Design Theory and Processes, Universidad Autónoma Metropolitana-Cuajimalpa, Mexico City, Mexico

Diana Paz FLACSO Ecuador, Quito, Ecuador

María Pozzio CONICET, Universidad Nacional Arturo Jauretche, Buenos Aires, Argentina

Laura E. Quiroz Rosas Department of Social Sciences, Universidad Autónoma Metropolitana-Cuajimalpa, Mexico City, Mexico

Pedro Roberto Jacobi Instituto de Energia E Ambiente—Universidade de São Paulo, São Paulo, Brasil

Mónica Velázquez Téllez Universidad Autonoma Metropolitana-Cuajimalpa, Mexico City, Mexico

Part I
Introduction

human societies in our days in strategic areas like health, energy, equality, etc. Many features of the current physical, political, social, and economic configurations have proven to be limited in resolving many of these issues. The systems of organizing housing, transport, or nutrition established in the last century show severe tensions like the excessive use of resources, the creation of climate-sensitive emissions, and other problems that have led to diminishing levels of sustainability, quality of life, and equality.

There is a vast production on the future of cities that can be traced at least back to the nineteenth century. The imaginary future of cities has been the subject of science fiction writers, architects, illustrators, filmmakers, videogame designers, fashion designers, restaurant chefs, etc. Alongside these visions of futures, urban studies have recently generated a series of future scenarios that are widely accepted by the expert community despite often narrow pathways to the future (Gonzalez-Arellano 2024). These "hegemonic" futures are meta-narratives that indicate not only how probable futures of cities are projected but also how to imagine diverse possible futures that are compared to specific selected preferable futures and specific trajectories to approach them. The established forms of institutionalization of city and urban futures are, in our opinion, dominated by a small group of guiding assumptions that can be synthesized into the following components:

- Toward a planetary urbanism: One of the main transformations of cities is updating the very conception of the city and the urban. For some authors, the category "city" is becoming obsolete in the face of the new conditions of urbanization, a process that has taken on an unprecedented dynamic of reconfiguration in recent years. It is not only a demographic and quantitative issue; "the urban" is surpassing the socio-spatial reality of the "city" to establish itself in the most diverse spaces of the planet: alpine resorts, transatlantic routes, deserts, even in and out of the terrestrial atmosphere (Brenner and Schmid 2014; Brenner 2018).
- Toward new geographies: One of the effects of this planetary urbanization is the generation of new geographies. Cities and "the urban" will take on new configurations, some in concentration dispersion, others in corridors or constellations of settlements or infrastructures. In addition to new urban forms, the future of "the urban" will be marked by new places, not only by the emergence of architectural or infrastructural innovations that will respond to new socio-cultural, political, and technological realities but through the adaptation of "old" places and structures of the built environment. Rethinking urban geography also means rethinking the coexistence between cities and nature as part of scenarios of "nature futures" (Mansur et al. 2022).
- Toward new temporalities: Future imaginaries of the city often emphasize socio-spatial transformations and ignore the other dimension of social reality: time. The city of the future will not only be polychronic (multiple temporalities), it will also experience a gradual temporal extension (24-h city), the acceleration of urban rhythms, and the synchronization of activities with distant places will be accompanied by new temporalities such as asynchronous and simultaneous ones (Madanipour 2017). If space were the dominant dimension in the industrial city of

the nineteenth and twentieth centuries, in the city of the future new temporalities (and chronotopes), would be the lever of innovations of the urban and the city (Zeiderman 2008).

- Toward new urban metabolisms: Urban crises associated with environmental problems mark the need to reconfigure urban metabolism. Current flows of resources and energy in cities are unsustainable, and the presence of specific "weak signals" already suggests a paradigm shift. Initiatives such as local production and consumption of food, clothing, water, energy, and raw materials are adopted in new urban lifestyles. The urban footprint that currently exceeds tens of thousands of kilometers from urbanites will, in the future, tend to be more selective and introduce aspects of social, economic, and ecological responsibility (Zhang et al. 2015). The transformation of sociotechnical and socio-ecological systems opens windows of opportunity for preferable futures, which is crucial for successful climate mitigation and adaptation (Egerer et al. 2021; McPhearson et al. 2016).
- Toward new forms of governance: The social production of the future of the urban and of cities requires an updating and reconfiguration of the governance of cities and territories. The way of imagining the future of cities is based on processes of vertical institutionalization of the future constituted by regulatory, normative, and cognitive instruments that often adopt path dependencies and, therefore, are not very innovative. The mismatch between the institutionalized governance of the city and a new socio-ecological reality characterized, among other things, by the aspects mentioned above will favor the emergence of innovations and new forms of participative governance (Sheikh et al. 2023).

All these dimensions of city futures can be associated with specific processes of creating images, narratives, or models of the future on different levels of individual and collective decision-making (UN-Habitat 2022). This process depends on diverse capacities to build and compare different futures in new ways. The existing forms of representation of the city of the future on a global level (like within the C40 group of cities for climate leadership, Davidson et al. 2019) or the images related to particular cities that are expressed, for example, in foresight exercises or strategic planning processes (Ravetz and Miles 2016), are often somewhat limited to linear, rigid extrapolations of existing structures, systems, and mental models. The reason for the persistence of narrow pathways to the future is the underlying paradigms and methods that are used to formulate narratives that reflect practices associated with the more traditional future studies and that rely heavily on quantitative techniques with anticipatory solid assumptions about the continuity of past and current configurations in the tradition of forecast models. Although the scope of the exploration of alternative futures has been slowly widened in recent years by using broader strategic foresight techniques to imagine alternative futures scenarios (Inayatullah 2011) and get a better understanding of the existential risks and opportunities, the potential of city governments and citizens to imagine and apply alternative futures is often rather limited.

Anticipation theory, one of the more recent approaches of contemporary futures studies, has developed a framework to characterize the anticipatory capacities of all kinds of social systems (Miller et al. 2018; Poli 2019). They part from the idea that social systems create images of themselves in the future that play an essential role in the diverse decision-making processes of all actors in the present. Even if the prevailing images of the future contain no changes, the system uses these images by orienting its resources and policies toward a reproduction of itself, avoiding variations. The creation and diffusion processes of future images express specific forms of "using the future" to guide processes of continuity and change. These processes are embedded in broader social dynamics that frame the construction of futures in urban contexts.

This book aims to characterize cities as evolving anticipatory systems with diverse implicit and explicit uses of images of the future. Each contribution explores specific aspects of anticipation in cities worldwide, particularly in the Global South, with recent contributions reflecting on the decolonization and construction of futures (Miraftab 2017). All the chapters of this book are related to specific aspects of anticipatory systems and some of their conceptual building blocks, like anticipatory assumptions, anticipatory capacities, or anticipatory governance; there is no homogeneous perspective or methodology behind the different efforts. The participating authors explore specific characteristics of cities as anticipatory systems in specific urban contexts, mostly in different parts of the Global South.

The guiding thread of this work is based on the postulate that the city, as a complex socio-spatial system, contains an inherent attribute, the capacity for anticipation. Cities and, even more so, metropolises are a set of coordinated systems at different scales. Urban studies have explored the analysis of cities from the perspective of systems theory from various approaches, for example, as socio-ecological systems, urban service systems (health, energy, transportation, etc.), physical systems (morphological of the built environment), socio-economic systems, political systems, etc. Each of these urban systems constructs its vision of the future that allows for sharing projects with other actors and actors of the various systems of the cities. This set of shared visions allows a set of values, regulations, and urban knowledge to be oriented toward the future, which is what we call the institutionalization of the future of cities (González-Arellano 2022). The institutionalization of the future of cities has a pragmatic component: anticipation. While the future scenarios of cities are collectively constructed and updated, it is through anticipation that these imaginaries are translated into action. As we will see throughout the chapters that make up this book, the anticipation of cities is embedded in each city system. In different ways, urban systems update their present state in the face of the different future scenarios they can imagine. This inherent attribute of anticipation of systems is diverse and uneven among urban systems, inhabitants, and places of cities.

This book is divided into four sections. Section I, "New Conceptual Frameworks for Cities as Anticipatory Systems," comprises two chapters that propose theoretical and conceptual frameworks for anticipatory systems in cities. Salomón González explores in this Chapter under the title "Cities as anticipatory systems" how anticipatory systems apply to cities, decomposing anticipation into future disposition, predictive capacity, and transformative updating. Due to their complexity and decentralized nature, cities harbor diverse anticipatory regimes, forming urban anticipatory communities shared among actors. Urban futures' production shapes policies rooted in diverse visions. Institutionalized futures highlight cities' anticipatory nature. In Chap. 2, titled "Hopeful Determination as an Urban Anticipatory System. An Enactive Approach," Blanca Miedes Ugarte and Manuela Á. Fernández-Borrero delve into urban development dynamics through a lens of hopeful determination. It intricately analyzes foresight, adaptability, citizen engagement, and multi-stakeholder cooperation, highlighting their role in fostering urban resilience and proactive change. Enactive cognition theory underpins the argument, emphasizing urban actors' active role in shaping their environment, thus underscoring the significance of "hopeful determination" in urban planning and development.

Section II of "Building Anticipatory Systems in Urban Sectors (contexts) in the Global South" includes four chapters that explore social configurations of anticipation processes in four different cities in the Global South on topics like mobility, care, health, and construction. In Chap. 3, titled "Sustainable Urban Mobility as an Anticipatory System: The Case of the Quito Metro," Marco Córdova and Diana Paz conceptualize sustainable urban mobility as an anticipatory mechanism within urban development. Future projections of socially inclusive, environmentally sustainable, and economically efficient urban configurations inform present policy and planning. This chapter analyzes through a qualitative case study the role of the metro system in Quito as a sociotechnical system that anticipates growth, climate change mitigation, and socio-cultural dynamics. Daniela Lossigio and Maria Pozzio wrote Chap. 4, titled "Sustainability of Life and Care in the Metropolitan Area of Buenos Aires, Argentine: Feminism and Anticipation." The chapter advances a discourse blending future studies and feminism, elucidating the slogan "*Se va a caer*" (Patriarchy will fall) to explore future narratives amid threats and the ascent of the ultra-right. The authors trace theoretical genealogies to conceptualize care and its anticipatory implications for practical proposals. In Chap. 5 ("Inclusive Urban Health Systems and Universal Well-Being: Mirage or a Real Future?"), Bikramaditya K. Choudhary relates the survival of humanity to the dependence on health systems for preventive and curative measures. Urbanization will require an understanding of urban complexities and livability. Future cities are deemed synonymous with "healthy" systems and economies, in plain contradiction to reigning neoliberal regimes. The concept of an "anticipatory city" is crucial. Despite visions for an unknown future, the paper addresses inequality and unchecked consumption in future cities. Mohammadjavad Mahdavinejad and Farzaneh Gharaati look at anticipatory processes related to construction paradigms in the Global South in Chap. 6, titled "Anticipating the future of Iranian cities: High-tech vs. nature-based solutions." In this chapter, the authors employed the discipline of anticipation and a co-design methodology involving experts in urban planning to envision future scenarios, drawing from future literacy approaches. Anticipated urban futures encompass high-technology, capital-intensive developments contrasted with nature-based, socio-economically empowering, and eco-friendly solutions. The study advocates for a future focused on happiness metrics, circular economies, and novel value creation, culminating in a proposed scenario for the Global South's urban future.

Section III comprises four chapters that analyze different levels and subsystems of cities as anticipatory systems in Mexico. In Chap. 7, Nora Morales and Laura Quiroz examine under the title “Anticipating asbestos-free futures—participatory mapping as elements to envision transitions based on hope” the co-creation of future visions within an asbestos-affected community on the outskirts of Mexico City, underscoring the significance of employing Transition Design principles in addressing complex social-ecological challenges. Through participatory mapping, community members generated alternative scenarios challenging normative urban futures. Expressions of hope related to preferable futures emerge as a motivational factor fueled by community narratives, driving action toward envisioned outcomes. Mónica Velázquez revises in Chap. 8 (“Anticipation in three restaurants of Mexico City: Imagining edible fungi in urban food futures”) how three mushroom-based restaurants in Mexico City anticipate urban food transitions amid the current eco-social crisis. It discusses international food scenarios, urban food dynamics, mushrooms’ emergence, their role in the Mexican food system, and the anticipatory capacities of these restaurants. Another contribution by José Luis González in Chap. 9 (“Anticipatory Systems and Security in Mexican Cities. A case-based review”) reflects on urban security anticipation, examining attempts by various urban actors and their influence on the city’s anticipatory capacity. Considering conflicting discourses and actions, it questions whether cities can constitute effective security anticipation systems. In Chap. 10, Bruno Gandlgruber and Marco Jaso explore zero-plastic futures under “Anticipating zero plastics futures in urban contexts in Mexico.” Based on a brief overview of the consequences of plastics contamination and some general characteristics of the plastics economy, they analyze different images, narratives, and models of the future that are related to the plastics futures and their applications to cities in the Global South. In the second part of the chapter, they give an overview of regulation processes related to plastics in Mexico to describe some elements of an anticipatory system in construction in this area.

Section IV includes anticipatory elements for urban climate policies. Pedro Roberto Jacobi, Marcos Tavares de Arruda Filho, and Rafael Calderon compare in Chap. 11 (“Recife and the anticipation of extreme weather events—paths to advance climate justice”) debates on climate change governance, especially in urban areas as crucial elements of public management. Anticipatory governance requires long-term vision changes to prevent crises. Recife, highly vulnerable to climate change, illustrates the importance of anticipation in local agendas. Finally, Brais Estévez, Julia Hidalgo, and Marion Bonhomme, in Chap. 12 with the title “Anticipatory action and future urban living in a context of increasing temperatures: an analysis from the Barcelona Climate Shelter Network” show how in recent years, global uncertainties, stemming from ecological and social crises linked to human-induced climate change, pose existential threats to cities and the planet. Local governments respond with anticipatory initiatives like Barcelona’s Network of Climate Shelters, engaging citizens in collaborative resilience-building efforts to address urgent climate challenges.

References

Brenner N (2018) New urban spaces: urban theory and the scale question. Oxford University Press, New York

Brenner N, Schmid C (2014) Planetary Urbanization. In: Brenner N (ed) Implosions/explosions: towards a study of planetary urbanization. Jovis, Berlin, pp 160–163

Davidson K, Coenen L, Gleeson B (2019) A decade of C40: research insights and agendas for city networks. Global Pol 10(4):697–708

Egerer M, Haase D, McPhearson T, Frantzeskaki N, Andersson E, Nagendra H, Ossola A (2021) Urban change as an untapped opportunity for climate adaptation. Npj Urban Sustainability 1(1):22

González-Arellano S (2022) The future scenarios of cities: an analysis of their institutional construction. Int J Urban Reg Res 46(5):865–873. https://doi.org/10.1111/1468-2427.13103

González-Arellano S (2024) Tipología de escenarios futuros de ciudades. Revista EURE - Revista De Estudios Urbano Regionales, 50(151). https://www.eure.cl/index.php/eure/article/view/4911

Inayatullah S (2011) City futures in transformation: emerging issues and case studies. Futures 43(7):654–661

Madanipour A (2017) Cities in time: temporary urbanism and the future of the city. Bloomsbury Publishing

Mansur AV, McDonald RI, Güneralp B, Kim H, de Oliveira JAP, Callaghan CT, Pereira HM (2022) Nature futures for the urban century: integrating multiple values into urban management. Environ Sci Policy 131:46–56

McPhearson T, Iwaniec DM, Bai X (2016) Positive visions for guiding urban transformations toward sustainable futures. Curr Opinion Environ Sustain 22:33–40

Miller R, Poli R, Rossel P (2018) The discipline of anticipation: foundations for futures literacy 1. Routledge, Transforming the future, pp 51–65

Miraftab F (2017) Insurgent practices and decolonization of future (s). The Routledge Handbook of Planning, Routledge, pp 276–288

Poli R (2019) Handbook of anticipation: theoretical and applied aspects of the use of future in decision making. Springer

Ravetz J, Miles ID (2016) Foresight in cities: on the possibility of a 'strategic urban intelligence.' Foresight 18(5):469–490

Sheikh H, Mitchell P, Foth M (2023) Reparative futures of smart urban governance: a speculative design approach for multispecies justice. Futures 154:103266. https://doi.uam.elogim.com, https://doi.org/10.1016/j.futures.2023.103266

United Nations Human Settlements Programme (2022) World cities report 2022: envisaging the future of cities. United Nations Research Institute for Social Development

Zeiderman A (2008) Cities of the future? Megacities and the space/time of urban modernity. Critical Plann 23–39

Zhang Y, Yang Z, Yu X (2015) Urban metabolism: a review of current knowledge and directions for future study. Environ Sci Technol 49(19):11247–11263. https://doi.org/10.1021/acs.est.5b03060

Part II
New Conceptual Frameworks for Cities as Anticipatory Systems

Chapter 2
Cities as Anticipatory Systems

Salomón González-Arellano

Abstract The production of urban futures significantly influences public policies and decision-making. Futures emerge from a diversity of visions that are continually updated. However, amid this diversity, many urban actors share an institutionalization of futures. The institutionalization of futures contributes in a certain way to the anticipation of cities understood as an inherent property of complex systems. This chapter aims to demonstrate how anticipatory systems can be applied to cities. This anticipatory property of cities is neither homogeneous nor static in the various subsystems that make up cities. In this chapter, the notion of anticipation is decomposed into two dimensions: (a) the future disposition and predictive capacity of the systems and (b) the transformative or updating capacity of the system as a response to possible evolution. Because of their complex nature and decentralized system, cities present a diversity of complementary future and anticipatory regimes, and the diversity of future and anticipatory regimes of cities are shared among urban actors, forming urban anticipatory communities.

Keywords Anticipatory systems · Urban systems · Future regime · Anticipatory communities · Urban modeling

2.1 Introduction

There is an international consensus that the world will be increasingly urban and that cities have gained relative influence as they face the central governments of the nation-state (Herschel and Newman 2017; Johnson 2018). This idea attributes a hegemony to cities due to their demographic and economic dimension and identifies them as strategic spaces for triggering innovation processes. Hence, cities are conceived as decision-making spaces and promoters of change while having a central place in producing imaginaries of futures (European Commission 2019; Ferraris et al. 2018).

S. González-Arellano (✉)
Department of Social Sciences, Universidad Autónoma Metropolitana-Cuajimalpa, Mexico City, Mexico
e-mail: sgonzalez@cua.uam.mx

S. Gonzalez-Arellano and B. Gandlgruber (eds.), *Cities as Anticipatory Systems*, The Urban Book Series, https://doi.org/10.1007/978-3-031-78162-9_2

The production of urban futures significantly influences public policy design and decision-making (van Dorsser et al. 2018). Recent literature points to the importance of anticipation in public policy conception and implementation. These are practices that some authors refer to as "long-term policies," as they are conceived for at least one human generation and handle deep uncertainty both for the generation of the problem they address, as well as for the conception of responses to it (Koskimaa et al. 2021; Meissner 2012; Organization for Economic Cooperation and Development [OECD] 2019; Sprinz 2012).

From a strategic rational perspective, the construction of future scenarios can be considered a tool that influences the development of cities. This capacity to decide why and how to use imagination to introduce the non-existent future into the present is known as Futures Literacy and consists of a key competence in the development of collective intelligence of cities and territories (González Arellano 2014; Miller 2018). Thus, scenario production plays a communicative role by focusing on strategic and sensitive issues and synthesizing diverse visions and knowledge. In addition to the practical utility based on strategic rationality, imagining the future touches dimensions of culture, science fiction, religion, and even superstition (MacKay and McKiernan 2018).

There are many visions of the future of cities. Futures emerge from a diversity of visions that are continually updated with the arrival and consolidation of events in the present, new information, or changes in values. However, amid this diversity, concrete futures are institutionalized, identifiable, and shared by many urban actors. The institutionalization of city futures is influenced by images of hegemonic futures that hold particular agendas, such as those related to climate change, the notion of smart cities, the UN Sustainable Development Goals (SDGs), or global events such as a pandemic, terrorism, or mega-events (Olympic Games, concerts, summits of transnational organizations).

The process of institutionalization of the future identifies three "pillars" that interact in the construction of future images of cities: (a) the regulatory pillar, which guides (forces) governments and other city actors to generate future scenarios, often through the creation of laws, international commitments, urban planning agenda; (b) the cultural-cognitive pillar, which contributes to the construction of images of the future through shared knowledge and information of the various dimensions of the city; and (c) the normative pillar, which, with the set of values and norms common among urban actors, guides the construction of what the futures of cities should be (González-Arellano 2022). The institutionalization of futures contributes in a certain way to the anticipation processes of cities by incorporating predictive and prospective models of itself or the external environment into the city's formal and informal functioning and coordination.

This chapter aims to demonstrate how the notion of anticipatory systems, first developed by Rosen (1985) and later by other authors (Poli 2010; Miller 2018), can be applied to cities. This proposal is based on the following postulates: (a) cities can be understood as systems with properties such as self-organization, bifurcation, and adaptation, along with others that characterize complex systems; (b) anticipatory theories consider that systems (biological and social) can modify today (update) based on a predictive model being this an inherent property of anticipation (Rosen et al.), (c) based on the above ideas, cities can be understood as anticipatory systems, (d) because of their complex nature, and because they are a decentralized system, cities present a diversity of complementary future and anticipatory regimes, and (e) the diversity of future and anticipatory regimes of cities are shared among urban actors, forming urban anticipatory communities.

In this sense, it is relevant to better understand cities as anticipatory systems, especially in periods of unprecedented socio-ecological transformations. The current socio-ecological transition offers signals (weak and strong) of future changes that should trigger anticipatory processes of a very diverse nature in the cities' lives (González-Arellano 2024). This chapter proposes that understanding cities as anticipatory systems can open new perspectives on the new nature of the urban world and the actors and institutions that produce the city.

2.2 The City as an Anticipatory System (AS)

This section discusses the foundational concepts of anticipation that allow us to propose the cities as anticipatory systems. First, we discuss how the city has been conceived from the theory of systems and, in particular, how it can be modeled as a complex system. In a second moment, Rosen's proposal presents the concept of anticipation, followed by other authors. This section explains the two dimensions of anticipation (future imagination and transformative agency) and the anticipatory system (AS) notion. Finally, it is shown how the general notion of an anticipatory system can be an intrinsic attribute of the city.

2.2.1 The City as a Complex System

The city has been analyzed from many perspectives: historical, geographical, sociological, economic, political, and ecological. Its multidimensional nature is the origin of the emergence of the interdisciplinary field of urban studies. Among the lengthy repertoire of perspectives for the study of cities, approaches based on systems theory stand out. For more than half a century, cities and regions have been studied from systems theory. Synthetically, this approach involves identifying elements or components of the city or a region, their interrelationships, and the functions-structure that makes it a system (Pumain 2018). Indeed, this approach presents at least two scales that have dominated the specialized literature on analyzing urban systems. The first deals with urban systems resulting from interrelationships between cities at varying scales, such as regional, national, or international. These pioneering studies developed a theoretical framework of urban systems, which involves viewing cities not as isolated entities but as interconnected elements within broader regional or global systems. Urban systems theory considers the flow of people, goods, and information between cities to see how these interactions contribute to the overall dynamics of urban areas (Berry 1964; Blumberg 1971).

The second approach comprises urban systems within cities, such as road systems, transportation, energy grid, city services, urban food systems, and health systems. There are numerous versions and types of models of these urban systems (For a review of the evolution and diversity of urban systems modeling, see Batty 2005, 2013). For example, Blumberg (1971) presents a vision of the city from a series of subsystems: subsystems of spaces and structures of land uses, subsystems of functional elements of the city in social and economic terms, a third subsystem is transportation, and finally a system of services such as water, electricity, hospitals.

With the developing understanding and modeling of urban systems, concepts from complex systems theories were gradually incorporated into the study of cities. The contributions of complex systems theories to city studies improved the understanding of city analysis more broadly. Hence, the city as a complex systems approach adopts a definition as a system "in which large networks of components with no central control and simple rules of operation give rise to complex collective behavior, sophisticated information processing, and adaptation via learning" (Mitchell 2009 cited in Fabrizio Li Vigni 2020).

From French geography and with the theoretical contributions of complex systems and the theory of self-organization, geographers proposed remarkable advances in modeling cities and regions. Pumain and Sander have promoted a school of spatial modeling since the late 1980s and, in particular, of the self-organizing capacity of cities. These works show an excellent example of applying complex systems concepts to explain the behavior and evolution of cities. With the development of the new quantitative geography and the development of computer science, a path from the spatial modeling of cities allowed significant advances in the theoretical and applied reflection on the city as a system. An outstanding exponent is Batty (2005, 2013) and his team from the Center for Advanced Spatial Analysis (CASA) in London, who have contributed to the reconceptualization of the city, emphasizing its dynamic, open, adaptive nature, networks, flows, interactions, scales, places, and demonstrating the relevance of understanding the city as a complex system.

Although there are different approaches to analyzing cities from the complex systems theories, it is possible to identify common attributes that allow us to understand cities as complex systems. First, cities must be conceived as systems of systems, in other words, as a set of components that create subsystems constituting networks through interconnections. Second, interconnections facilitate flows of energy, matter, information, people, goods, and money, creating feedback loops that are more or less intense depending on the dynamism of places at different scales and distances. Third, this process is a property of dynamic and adaptive systems, giving cities emergent properties that translate into innovations of many different kinds: cultural, technological, social, etc.

City Components and City Subsystems. A city's physical infrastructure comprises various tangible structures, encompassing buildings, roads, bridges, utilities, parks, and other elements that constitute the city's physical environment. The human population residing and working within the urban area represents a pivotal component, exerting significant influence over the city's cultural milieu, economic landscape, and social dynamics (Pumain 2018). The economic systems inherent to cities are characterized by intricate networks encompassing businesses, industries, and employment opportunities, collectively contributing to the overall economic vitality of the urban entity. Concurrently, social systems within cities are shaped by the complex interplay of relationships, community structures, and cultural dynamics, collectively defining the city's social fabric and exerting substantial influence over individual behaviors and interactions among residents.

Urban Interconnections and Networks. The transportation networks within urban areas encompass a myriad of infrastructural components. This infrastructure plays a pivotal role in facilitating the seamless movement of individuals and enhancing accessibility throughout the urban expanse (Urry 2007). In tandem with physical transportation networks, cities are intrinsically linked by communication networks, constituted by digital infrastructures such as the Internet and telecommunications. These technological frameworks play a critical role in fostering the exchange within the city itself and extending beyond its geographical confines (Batty 2013).

Furthermore, the economic vitality of cities is intricately tied to supply chains, wherein the flow of goods and services assumes paramount significance. Cities rely extensively on the inflow of commodities from external sources, necessitating the establishment and maintenance of complex supply chain networks to meet the diverse needs of their populace. This dynamic interdependence underscores the multifaceted nature of urban systems and the integral role of urban networks in sustaining the functional coherence of contemporary cities.

City Feedback Loops. Economic feedback mechanisms within urban contexts denote the reciprocal relationship between alterations in economic conditions and ensuing effects on employment rates, thereby exerting consequential influence over consumer spending patterns and business activities. Fluctuations in economic variables contribute to a dynamic interplay that shapes the intricate network of economic interactions within the urban milieu. Similarly, environmental feedback processes delineate the reciprocal interactions between urban development activities and their impact on the local environment. The alterations brought about by urbanization can instigate environmental changes, thereby influencing the overall well-being of city residents. The intricate interdependence between urban development and environmental dynamics underscores the need for a comprehensive understanding of these feedback loops that help inform sustainable urban planning practices and policies prioritizing the holistic welfare of the urban populace and the surrounding environment.

Cities Emergent Properties. A distinctive amalgamation of individuals, traditions, and historical narratives constitutes cultural identity within an urban context. This unique cultural identity of a city materializes through the dynamic interactions among diverse individuals and communities. The resulting synthesis of these components contributes to forming a city's distinct cultural character, encapsulating the shared values, practices, and expressions that collectively define its identity. Furthermore, cities commonly function as focal points for innovation, fostering an environment where diverse ideas and talents converge. This convergence catalyzes the generation of novel technologies, artistic endeavors, and cultural trends. The intricate interplay of varied perspectives and skill sets within urban settings engenders a fertile ground for creativity to flourish. Therefore, cities emerge as dynamic hubs wherein synthesizing diverse intellectual and creative inputs engenders a continuous innovation cycle, influencing technological advancements and artistic and cultural expressions.

City Adaptability. Urban planning plays a pivotal role in the adaptive evolution of cities, as it involves formulating and implementing strategic development approaches to contend with prevailing challenges while proactively anticipating future needs. This process encompasses a comprehensive examination of the urban landscape, considering population growth, economic dynamics, and environmental sustainability to inform judicious planning and development initiatives. Simultaneously, the resilience of cities assumes paramount importance in the face of multifaceted challenges. Urban environments are susceptible to shocks, encompassing economic downturns, natural disasters, and public health crises. Navigating such adversities necessitates establishing adaptive systems and formulating contingency plans to mitigate the impact of disruptive events. The imperative for resilience underscores the significance of strategic foresight, robust infrastructure, and proactive governance in fortifying cities against unforeseen challenges and ensuring their sustained functionality and well-being.

Understanding cities as systems provides a holistic perspective that can inform urban planning, policy-making, and sustainable development efforts. It acknowledges the interconnectedness of various elements and the need for coordinated efforts to create and maintain a thriving urban environment.

2.3 Anticipatory Systems (AS)

As discussed in the introductory chapter, future studies have devoted relevant attention to anticipation. The relevance of recent contributions from the anticipation studies community points to the gradual consolidation of anticipation as a discipline (DoA) in its own right (Miller 2018). An excellent example of this dynamic field is the two volumes of the collective work Handbook of Anticipation, Theoretical and Applied Aspects of the Use of Future in Decision Making (Poli 2019) to realize the broad spectrum that anticipation theories offer as a legitimate research topic.

In a highly synthetic way, anticipation is understood as a process when "the future is used in action." In other words, anticipatory processes allow the translation of future scenarios or images into present behaviors. This process radically differs from reaction, which implements action after an event. A reaction should be seen as a complementary process to anticipation. Anticipatory processes can respond to predictive models outside or inside the system (Poli 2019). For example, a city can trigger an anticipatory process in the face of dangerous meteorological scenarios, such as launching preventive actions in the face of a hurricane whose extrapolation of its trajectory represents a threat. Alternatively, anticipatory actions can be developed in the face of demographic scenarios of its population, such as the creation of care policies or pension funds in the face of an aging population. From this approach, anticipation is composed of two dimensions: (a) a willingness and ability to imagine future scenarios, and (b) integrating this vision of the future to take some action.

The distinction between these two dimensions is important because, from this perspective, predictive or prospective exercises are not, strictly speaking, a form of anticipation since they are limited to generating visions of the future and do not lead to action in the present. As mentioned by Poli, these two dimensions of anticipation are consistent with Rosen's definition: "An anticipatory system is a system containing a predictive model of itself and its environment, which allows the system to change state at an instant in accord with the model's predictions about a later instant" (Rosen 2012:8, 313, originally published in 1985).

Considering the above, we can synthetically construct an anticipatory system for the city from other anticipatory system modeling exercises. Suppose we have the system of a particular city that we can consider as system S. City S can be understood as a dynamic system for which we can make predictions of states of S based on the knowledge of an initial state and input of the system. For example, a city's food system is dynamic with some degree of behavioral knowledge: production, distribution, sales, consumption, and waste. These are known with some degree of confidence.

However, let us explicitly incorporate a predictive model of city S into another system we will call M. As a dynamic system, the city system S evolves in real-time. At the same time, the possible trajectories for the predictive model M follow a time variable faster than the real-time of S. Therefore, any change observable in the model of M serves as a predictor of the behavior of some observable attributes of the city system S at a later instant. Regarding the example of the urban food system, the predictive model of system M will be able to generate trajectories of changes in the food supply due to a prolonged drought or war thousands of kilometers from the city system S (see Fig. 2.1).

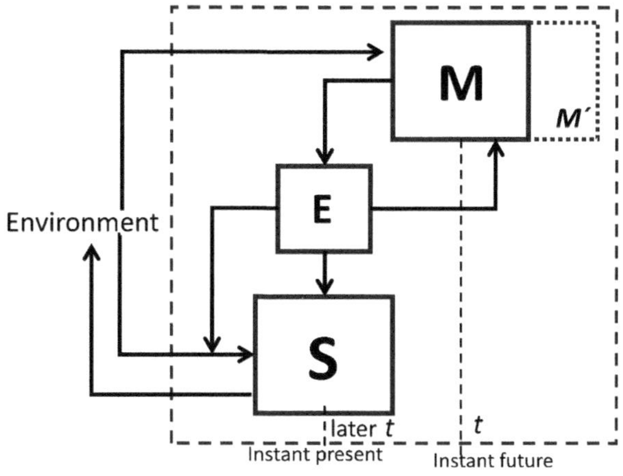

Fig. 2.1 Anticipatory system (adapted from Rosen 1985)

Table 2.1 Future regimen adapted from Chateauraynaud and Debaz (2017)

	Time modeling	Logic of action
High-frequency	Continuous flow	Visual navigation; uncertain anticipation in a very short time
Urgency	Time is short	Race against time to save a situation
Expectation	Suspended time	Continual displacement of expectation time
Optimization	Adaptation to an open future	Automatization of the learning process; creative randomness
Anticipation	Accelerated time	Action over an ongoing process; preparedness
Prediction	Calculated and linearized time	Calculation space planning
Prospective	Time is non-linear and open	Scenario production
Promise	Delay logic	Engaging in a promise, proving one's credibility
Prophecy	Eschatological time	Announcing an inevitable future

available to transform both the system and the environment, (d) the diversity and quantity of material and immaterial resources available.

From the intersection of these two dimensions, it is possible to identify diverse ways of practicing anticipation at the individual and collective levels. That allows us to use the notion of anticipatory communities developed by Li Ving. For this chapter, we understand as anticipatory communities the group of actors who share a specific anticipatory practice, i.e., they share both a future regime and anticipatory policies as their transformative capacity (i.e., a prescriptive vision of the environmental governance we need to implement). Therefore, communities of anticipation share a vision of the future, a prescriptive ideology, and a practical investment in a city and governance project with which to face the future.

We can expect diverse actors, lifestyles, interests, functions, and interactions in a complex social system like cities. Amid this diversity, there are affinities of all kinds: political, professional, ideological, religious, residential, food, sports, clothing, music, and others. These affinities can take the form of formally or informally recognized communities. In addition to this diversity, the actors have unequal transformative capacity to transform their system and influence their environment.

Figure 2.2 shows an adaptation of the proposed future regimes grouped by their similarity in interweaving the two dimensions of anticipation: (a) temporal dimension, frequency of change patterns, and (b) transformative capacity or updating of the system. As shown in the figure, it is possible to identify five groups corresponding to the relationship between the two dimensions of anticipation: future regimes and transformative capacity to update the present.

The first group of anticipatory practices (C1) responds to high-frequency, random, with little time of anticipation, and highly changeable (or with the flow of time stopped in the case of expectation) situations. Here, we find the predictive modes

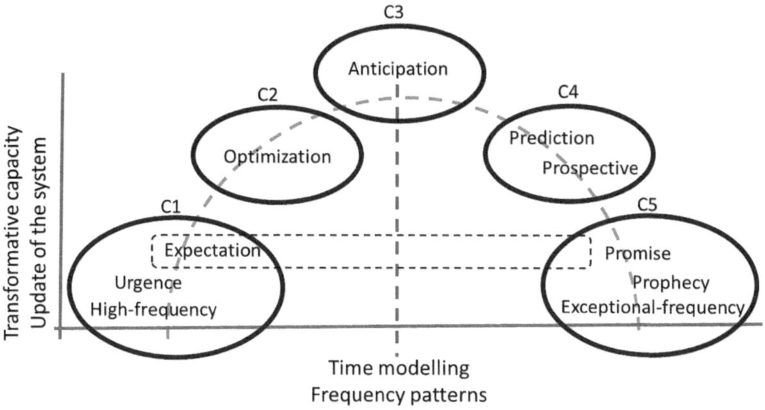

Fig. 2.2 Future regimes and transformative capacity of the system. *Source* Prepared by the author

of high-frequency events, urgency, and expectation, which urban actors treat with a low transformative capacity of the system. This group of anticipatory practices is very close to reactive systems because they are very close in time to events (almost "too late"). In this group, as in C4, we have included the expectation of maintaining a poorly defined short and long temporal range.

The practices we have named optimization (C2) start from a future regime with high-frequency and regularity event patterns, which allows a high degree of predictability. Actors adopting optimization as anticipatory practices modify the system, relying on this high degree of predictability to maximize benefits and reduce costs. It is a regular process of learning and adaptation with a low degree of transformative change aimed at improving the current functioning of the system.

Anticipation (C3) is the ideal way of updating the system in response to a predictive model of the system or its environment. It is more potent than optimization, prediction, and foresight because it adopts a greater transformative capacity and updates the state and functioning of the system. It is similar to prediction and foresight in generating scenarios in that it is less interested in contingency and more interested in medium- and long-term scenarios and is open to alternative scenarios. This practice is the one that, in the strictest sense, corresponds to an anticipatory system.

The prediction and prospective practices of the group (C4) often adopt a medium and long-term future regime. The objective is to elaborate scenarios in the realm of the possible under a linear or tendential vision for prediction and a non-linear and more creative, alternative, and exploratory one for prospective. In both cases, their primary purpose is to stimulate the imagination and is less about acting on the system's present state. Although they are inputs for anticipation, their primary purpose is not to update the present.[1]

Finally, promise and prophecy (C5) are, in the extreme, not very anticipatory practices because they mobilize few actors due to their distant time horizon. Some of these visions of futures go beyond the present generations and move into the realm of utopia, collapse, or, at best, hope. In this group, as in C1, we can include the expectation of maintaining a poorly defined temporal range.

2.5 Anticipatory Communities

We can understand an anticipatory community as a collective of actors that share (a) a future and anticipatory regime, (b) standard causal models of how the world works, and (c) shared future policies. From the above, it is possible to identify four types of anticipatory communities of cities: (a) the Repairers, (b) the Managers, (c) Anticipators-Strategists, the Visionaries-Counselors, and the Utopians-Prophets.

2.5.1 Repairers

This community corresponds well to the future and anticipatory regime C1 in Fig. 2.1 Urban actors with care and alert functions (anticipation) in the face of high-frequency events, emergencies, and the expectation of events with very low predictability. This type of anticipatory community group has actors who have identified recurring patterns of high, random frequency and have little time to anticipate. Because of the quasi-random or sudden nature, the causality of events is often known, although the actors in the system have little control. Security, civil protection, and surveillance systems adopt this anticipatory mode. Part of the anticipatory actions of urban traffic management work on "sudden events with little reaction time."

The anticipatory policies of this community are characterized by the design and practice of protocols that allow, through protocols, the automatization of a series of appropriate behaviors for events (accidents) that follow familiar patterns. These communities are in many urban systems, in charge of governmental and non-governmental actors. Earthquake protocols have been an established practice in Mexico City for several decades. Unfortunately, in recent years, in some cities, the practice of protocols against attacks by organized crime or terrorist actors has been established.

[1] Although constructing visions of the future can transform present consciousnesses, these approaches do not, in their pure version, aim to transform the functioning and present state of the system.

2.5.2 Managers

This group of urban actors forms a community that is a suitable adopter of the anticipatory regime based on optimization practices. Like the repairers, they do not seek a transformative change in the state or functioning of the system but rather an incremental improvement based on learning predictable behavioral patterns. Their relationship with the future is usually short and medium-term, with a reliable internal predictive model and low uncertainty. Indeed, unlike the group of repairers, managers focus on anticipation in the face of predictive models internal to the system, and therein lies their most excellent agency in establishing modest changes to improve the current performance of system functions.

The anticipatory policies of this community can come from strategic management practices, such as continuous improvement. Some examples of this can be urban systems such as transportation, energy, or food waste management, which, based on regular patterns and low uncertainty, try to maximize benefits and minimize costs, such as the optimal design of service routes and energy consumption schedules. This anticipatory community can be found in the governmental sector, companies, and social and political organizations whose logic of maximizing benefits differs from the economic one and translates into social or political impact with the least effort.

2.5.3 Anticipators-Strategists

The community of anticipators is composed of urban actors with a high level of agency, capable of influencing the state and functioning of the system based on internal or external predictive models. They are actors with a good understanding of the complexity of urban systems, able to identify relevant interactions between components, redundancies, and bifurcations. Actors of this anticipatory community share more complex causal models than the previous groups, are familiar with medium and long-term future regimes, have a creative and open perspective on predictive models of the future of cities, and are alert to interactions with the environment as external to the system.

The anticipatory policies of this community are characterized by adopting both a pragmatic and creative stance toward the construction of future scenarios often related to relevant problems such as climate change, food, population aging, and others. Predictive models of these problems require putting transformative actions beyond improving (optimizing) the current state and functioning of the systems. For this reason, these communities will generally be outside the government or organizations' management routine (managers). Some of the profiles of the members of these communities are artists, creatives, activists, consultants, politicians, entrepreneurs, or academics. We can distinguish three types of anticipators: (a) internal anticipation, oriented to prepare for the predictive model of the future; (b) deep anticipation, the transformers of the future scenario; and (c) open anticipation, which is the anticipation of the future.

2.5.4 Visionary-Advisors

This group of actors share future visions and anticipatory regimes based on prediction and prospection. In urban systems, these communities often present themselves as experts capable of constructing medium- and long-term images of a parcel in urban reality. They often share techno-scientific rhetoric for constructing medium and long-term scenarios based on rational, objective, and more or less transparent methods.

Their anticipatory practices are limited to the construction of predictive models that offer clarity of the future. The members of this anticipatory community may be close to or distant from decision-makers and may present a diversity of methods and very diverse topics. From forecasting a linear and trend-based vision of the future, we find actors who model the territorial growth of cities, demographic scenarios, and even some climate scenarios. On the other hand, prospection is also oriented toward creating scenarios, but in a less deterministic and trendy way than prediction. Emphasis is placed on the diversity of alternative futures.

This community is essential as it represents many actors involved in the institutional construction of city futures. Although, in principle, they do not seek to modify the current state and functioning of the system since their function is to generate future scenarios. Their influence lies in the production of futures. Institutional processes, such as producing urban development plans, demographic or climate scenarios for programs or laws, and more, often endorse it. In addition, this group can contribute to the Anticipator-strategist community by providing inputs to create their predictive models.

2.5.5 Utopian-Prophets

At the extreme side of the dimension for time modeling are actors who share concerns about the long and very long-term future. This anticipatory community comprises visions of the future that fall into utopia or collapse. Some scenarios surpass the present generation's horizon, the city's existence, or even humanity. In this group, we also include anticipatory practices of the "promise" or "hope" type, in the sense that they share a mobilizing force and expectation that presents a low transforming agency of the current state and functioning of the system.

Their anticipatory policies are characterized by low or no agency over the system's present state. Their purpose is to warn, orient, and make the value systems of the current social and ecological system visible. In these communities, hopeful, utopian, dystopian, and catastrophic positions can be observed simultaneously. In this community, we can find actors of diverse profiles, from artists, creative people, science fiction writers, religious people, philosophers, and activists.

2.6 Anticipatory Operators

These typologies of anticipatory communities are ideal models; in the real life of cities, the actors share more than one community, and some of the anticipatory practices are superposed and adapted according to the situation and the personal and institutional resources available. None of these communities is better or worse than the others, as they are all products of culture and the shared futures literacy in the incessant production of cities.

So far, the examples of anticipatory systems in cities suggest a process focused on actors, i.e., on anticipatory actions performed by individuals or organizations. It is possible to broaden the notion of an anticipatory system by adopting a broader system view, going beyond human actors. Actor-network theory (ANT) allows, at least in a suggestive way, to imagine an anticipatory system of (human and non-human) actors. ANT approach emphasizes the importance of non-human entities, or "actants," in shaping social interactions and networks. This approach can help model anticipatory systems, especially because, as demonstrated by Rosen (1985) and numerous examples (Poli 2019), anticipation is an inherent attribute of both human and non-human systems. Lussault (2007) adopts the notion of actant and the main postulates of ANT to reflect on the role of spatial entities (human and non-human) in organizing societies. Lussault replaces the term actant with that of the spatial operator and, similarly to ANT, stable human and non-human spatial operators as actants with a spatial agency.

Therefore, following Lussault and ANT, anticipatory operators can be human and non-human actors that maintain a network of interactions in an anticipatory system. From this point of view, an anticipatory system of cities can incorporate the notion of anticipatory operators formed by persons and organizations, and other types of operators such as non-human living organisms (animals and plants), geophysical operators such as the weather, or active volcano, immaterial operators such as laws, information, values, or collective beliefs about the future. Human and non-human anticipatory operators establish networks by interactions, and in the case of cities, these networks are mainly decentralized and multi-scale (neighborhood, city, region, country, world).

The anticipatory function of cities is a distributed property, not centralized in a subsystem specialized in that function. However, it can be observed that in some cities, both in the government bureaucracy and in the organization of companies, departments, or teams formally dedicated to foresight, anticipation is an inherent attribute in each part of the system. That allows anticipatory practices to be developed at any scale of the system, to a greater or lesser extent, and from different approaches.

As already mentioned in another text, the institutionalization of the futures of cities is built from regulatory, normative, and cognitive arrangements (González Arellano 2022). We consider that the above is valid for forming anticipatory regimes and communities. This process establishes a hegemonic future that imposes conservative imaginaries and a narrow margin for imagining scenarios of cities "outside the box." That takes on even greater importance in a socio-ecological transition phase characterized by disruptive regime shifts (energy, climate, food, geopolitical) that occur with high uncertainty, complexity, and innovation. Faced with this unprecedented scenario, we fear that traditional planning and future construction methods are becoming less and less adequate.

2.7 Conclusion

As discussed above, the central idea of this chapter is to understand anticipation as an inherent attribute of socio-ecological systems, that is, as a property intrinsic to the functioning of systems. In this sense, and as developed in the first section of this article, if cities are understood as complex systems, then anticipation is a functional property of cities.

Cities as complex systems have the inherent property of anticipation. This anticipatory property of cities is neither homogeneous nor static in the various subsystems that make up cities. This chapter has reviewed the notion of anticipation, starting from two dimensions: (a) the future disposition and predictive capacity of the systems and (b) the transformative or updating capacity of the system as a response to possible evolution.

Following Rosen's founding proposal, we have adapted the general model of an anticipatory system for the city system in this chapter. This proposal was based on three aspects: (a) cities are complex systems, (b) anticipation is an inherent attribute of complex systems, and (c) therefore, cities present diverse anticipatory policies corresponding to the varied future regimes and the unequal transformative capacity of urban actors. In this chapter, we identified five modalities explained by the future regime adopted and the transformative agency of the system's actors: repairers, managers, anticipators-strategists, visionary advisors, and utopian prophets. These anticipatory types are not mutually exclusive nor exclusive to human actors; non-human actors also participate in the anticipatory systems of cities.

The anticipatory policies of cities are the result of the institutionalization of the future that cities develop. As mentioned, besides being diverse, anticipatory practices follow differentiated institutionalization processes. On the one hand, we can find formal or informal regulations that reward or punish certain actors for developing anticipatory processes based on very preset models or scenarios. An excellent example of this is laws and other types of incentives at all scales that address global climate change scenarios. In addition to regulatory instruments that incentivize cities' anticipatory practices, certain actors also share the normative dimension of what the city should be.

An example is the Charter for the Right to the City signed by Mexico City, which constitutes the collective construction of a city model to which an important group of actors adhere. This model of the future of the city takes the form of anticipatory practice in concrete and performative actors, such as the signing of the charter by the government of Mexico City's government in a public act and accompanied by stakeholders of the city, which had even influenced the drafting of the first Political Constitution of Mexico City in 2017. In addition to the regulatory and normative dimensions, a third dimension contributes to institutionalizing the future and anticipatory practices in cities. The cognitive and cultural dimension is a source of incentives for actors to build anticipatory models and anticipatory actions in cities. Access to information on the internal urban systems or the external environment of cities can trigger anticipatory initiatives. A recent and familiar example for us is the global pandemic of COVID-19. The information about the pandemic that quickly circulated the world allowed many governmental and non-governmental actors to take various anticipatory actions. A great example is governments initiating negotiations with pharmaceutical companies to purchase vaccines for the public and private sectors.

These three dimensions, regulatory, normative, and cognitive, interact and together contribute to the construction of coordination, conventions, and agreements, formal and informal, between actors on future and anticipatory regimes, more or less consensual between local or external urban actors. Here, such anticipatory regimes and communities are operating at various scales. International humanitarian aid can be an example of this.

Finally, the understanding and development of anticipatory systems often involves a combination of technological innovation, data-driven decision-making, collaboration, and a commitment to addressing specific challenges or needs. Public institutions, private organizations, and communities may employ different strategies and approaches to create and implement these systems depending on the context and goals. Comparing urban systems as anticipatory systems involves assessing how different cities or urban areas use future scenarios and transformative capacities supported by data, technology, institutional, and planning strategies to anticipate and address various challenges and opportunities.

References

Batty M (2005) Cities and complexity: understanding cities with cellular automata, agent-based models, and fractals. The MIT Press, Cambridge, MA, p 565

Batty M (2013) The new science of cities. The MIT Press, Cambridge, MA

Berry BJL (1964) Cities as systems within systems of cities. Pap Reg Sci Assoc 13:147–163

Blumberg DF (1971) The city as a system. Simulation 17(4):155–167. https://doi.org/10.1177/003754977101700404

Chateauraynaud F, Debaz C (2017) Aux bords de l'irréversible. Sociologie pragmatique des transformations. Éditions PETRA

European Commission (2019) Cities as innovation hubs. European Commission, the future of cities: opportunities, challenges and the way forward. Publications Office, pp 105–112. https://doi.org/10.2760/375209
Ferraris A, Santoro G, Papa A (2018) The cities of the future: hybrid alliances for open innovation projects. Futures 103:51–60. https://doi.org/10.1016/j.futures.2018.03.012
González-Arellano S (2022) The future scenarios of cities: an analysis of their institutional construction. Int J Urban Reg Res 46(5):865–873. https://doi.org/10.1111/1468-2427.13103
González-Arellano S (2014) Inteligencia territorial y la observación colectiva territorial. Espacialidades 04(02):89–109. http://espacialidades.cua.uam.mx/vol/04/2014/02/04_Gonzalez.php
González-Arellano S (2024) Tipología de escenarios futuros de ciudades. Revista EURE—Revista De Estudios Urbano Regionales 50(151). Recuperado a partir de https://www.eure.cl/index.php/eure/article/view/4911
Herrschel T, Newman P (2017) Cities as international actors. Urban and regional governance beyond the Nation State. Palgrave Macmillan UK. https://doi.org/10.1057/978-1-137-39617-4
Johnson C (2018) The power of cities in global climate politics. Saviors, supplicants, or agents of change? Palgrave Macmillan London/City Leadership Laboratory, University College London (UCL). https://doi.org/10.1057/978-1-137-59469-3
Koskimaa V, Rapeli L, Hiedanpää J (2021) Governing through strategies: How does Finland sustain a future-oriented environmental policy for the long term? Futures 125:102667. https://doi.org/10.1016/j.futures.2020.102667
Li Vigni F (2020) Five anticipation communities in complex systems sciences: complexity science and its visions of the future. Futures120:102551. https://doi.org/10.1016/J.FUTURES.2020.102551
Lussault M (2007) L´Homme spatial : la construction sociale de l'espace humain. Seuil
MacKay RB, McKiernan P (2018) Scenario thinking: a historical evolution of strategic foresight. Cambridge University Press
Meissner D (2012) Results and impact of national Foresight-studies. Futures 44(10):905–913. https://doi.org/10.1016/j.futures.2012.07.010
Miller R (2018) The discipline of anticipation. Exploring key issues. In: Miller R (ed) Transforming the future: anticipation in the 21st century. UNESCO, Routledge
Mitchell M (2009) Complexity. A guided tour. Oxford and New York, Oxford, University Press
Organization for Economic Cooperation and Development (OECD) (2019) Strategic foresight for better policies. https://www.oecd.org/strategic-foresight/ourwork/Strategic/Foresight/for/Better/Policies.pdf
Poli R (2010) The many aspects of anticipation. Foresight 12(3):7–17
Poli R (2019) Handbook of anticipation. Theoretical and applied aspects of the use in decision making. Springer
Pumain D (2018) An evolutionary theory of urban systems. In: Rozenblat C, Pumain D, Velasquez E (eds) International and transnational perspectives on urban systems, Springer
Rosen R (1985) Anticipatory systems. Philosophical, mathematical, and methodological foundations. Oxford, Pergamon Press (2nd edn Springer 2012)
Rosen R (2012) Anticipatory systems philosophical mathematical and methodological foundations. Springer, New York
Sprinz DF (2012) Long-term environmental policy: challenges for research. J Environ Dev 21(1):67–70. https://doi.org/10.1177/1070496511435667
Urry J (2007) Mobilities. Polity
Van Dorsser C, Walker WE, Taneja P, Marchau VAWJ (2018) Improving the link between the futures field and policymaking. Futures 104:75–84. https://doi.org/10.1016/j.futures.2018.05.004

Chapter 3
"Hopeful Determination" as an Urban Anticipatory System—An Enactive Approach

Blanca Miedes-Ugarte and Manuela Á. Fernández-Borrero

Abstract This chapter offers a compelling exploration into the dynamics of urban development through the lens of hopeful determination. This concept is illuminated through the case study of Plan Integral del Distrito 5 de Huelva (PID5H), showcasing how urban communities, equipped with a vision for the future, navigate and shape their developmental trajectories. The chapter intricately examines the interplay of foresight, adaptability, citizen engagement, and multi-stakeholder cooperation in fostering urban resilience and proactive change. It underscores the transformative power of shared narratives in rallying collective efforts toward inclusive and sustainable urban futures. By anchoring its argument in enactive cognition theory, the chapter argues that urban actors do not merely react to their environment but actively participate in its creation and evolution, embodying hopeful determination in their approach to urban challenges. This innovative perspective offers a nuanced understanding of how urban areas can anticipate and adapt to change, making hopeful determination a vital component of urban planning and development.

Keywords Hopeful determination · Enactive urbanism · Resilient cities · Participatory development · Anticipatory urban systems

3.1 Introduction

In this day and age, cities are facing challenges like never before. These challenges include socio-economic issues, such as the pursuit of prosperity by ensuring decent living conditions for the entire population and integrating different ways of life. They also have environmental challenges around managing urban metabolism, the effects of climate change, pollution, and biodiversity loss. Cities also face rapid

B. Miedes-Ugarte (✉) · M. Á. Fernández-Borrero
COIDESO-Universidad de Huelva, Huelva, Spain
e-mail: miedes@uhu.es

M. Á. Fernández-Borrero
e-mail: manuela.fernandez@dstso.uhu.es

S. Gonzalez-Arellano and B. Gandlgruber (eds.), *Cities as Anticipatory Systems*, The Urban Book Series, https://doi.org/10.1007/978-3-031-78162-9_3

cultural and technological changes that affect how people think, behave, and live together. These tensions are most acutely felt in the most deprived urban areas. Against this backdrop, there is a need for approaches to urban design that address existing problems and anticipate and prepare cities for an uncertain future. Here is where the concept of "hopeful determination" as an urban anticipatory system comes into play. The authors conducted an ethnographic study of the *Plan Integral del Distrito 5 de Huelva* (PID5H), a multi-actor, multi-level project initiated in 2000 to regenerate a deprived urban area of approximately 15,000 inhabitants in southwestern Spain. Through the lens of the enactive cognition approach, this chapter explores how the combination of long-term vision, adaptability, active citizen participation, multi-actor collaboration, and a multi-dimensional approach can build cities that not only confront but proactively anticipate and transform themselves in the face of emerging challenges, provided that actors are willing and able to at least partially articulate their visions of the future.

The chapter is divided into four sections. It introduces the enactive approach (Varela et al. 2017) as the basic conceptual framework, followed by discussing the concept of "hopeful determination" in the urban context using the three-horizon model (Sharpe 2013; Sharpe et al. 2016). The authors, coordinators of the Plan's scientific and technical team for over a decade, examine the evolution over time of the future horizons of the actors involved, as well as the conflicts that arose and the commitments made during the different phases of the Plan. This section also discusses the competencies developed by those involved in regenerating the urban space and co-constructing a more inclusive and resilient future.

More than a mere methodology or strategy, "hopeful determination" is a praxis, an integrated way of being, doing, and knowing in which actors and environment mutually model each other in an ongoing sense-making process in which mobilizing narratives play a defining role (Liveley 2017). The concept calls for engagement and pro-activity in urban design and reinvention. By recognizing the link between urban challenges and the inherent capacity of communities to innovate and adapt, this chapter explores the opportunities and constraints on the ground for building more resilient, inclusive, and sustainable cities.

3.2 An Enactive Approach to "Hopeful Determination"

The world can be considered a "place for things" or a "field for action." In the former, modern societies took a scientific approach to interpreting the world to define these things' verifiable and agreed-upon properties. We also explore how to use them efficiently for various purposes where appropriate. However, the world as a forum for action—an older view that includes the world as a place for things—is a place of "value, a place where everything makes sense," a sense that tells us who we are, that is shaped in social interaction, and that serves as an interpretive framework for action (Peterson 2002). Knowing what something is involves recognizing its particular attributes and motivational significance, i.e., what this means for its behavior.

That "something" must evoke an emotional response to garner enough interest for examination and analysis. Scientific methods try to remove affect from perception so that experience can be described universally. They are, therefore, woefully unsuited to the task of guiding action involving those all-pervasive emotional aspects of the human process.

The future can also be seen from the point of view of things to be controlled or instrumentalized. What we forecast or foresee will happen (predict) in the probable, plausible (explored) future. It can also be seen in terms of action, what we want to happen, or what we think should happen—desirable, normative, and prospective futures (Miller et al. 2013). All our ideas about the future, which we are more or less aware of, are a yardstick by which we judge the present. The envisioned, desired, and imagined future is a guiding framework for evaluating the present and informing action in the here and now, whether reacting to whatever is unfolding or taking transformative action to create new life possibilities. Our way of being in the world is anticipatory.

Understanding how we come to know is fundamental to understanding our relationship to the future. Recent advances in neuroscience suggest that the human way of making sense of the world is inherently predictive—indeed, it may be the brain's primary mode of operation. In a recent exploration of the science of consciousness, Seth (2021) posits that consciousness is a hallucination of the brain, akin to a controlled dream in which the brain makes its best prediction about what is going on in the outside world in response to incomplete and ambiguous sensory cues. That implies that our conscious experience of the world is a personalized internal simulation and that our experience of "reality" is an interpretation crafted by the brain. The science of emotion further explores this pivotal role of prediction in human existence (Barrett 2017). Emotions are presented as predictions that prime the body to act in line with learned expectations and environmental demands and are heavily influenced by cultural context and social norms. Concepts and language help us to construct this social reality.

In this vein, O'Hara and Leicester (2012) include such *epistemic literacy* among the three critical competencies required for navigating the VUCA (vulnerable, uncertain, complex, and ambiguous) landscape of the twenty-first century, along with an understanding of one's agency (*psychological literacy*), and the ability to read and facilitate cultural processes (*cultural literacy*).

The enactive approach to cognition is an approach that emphasizes the vital role of prediction in cognition, offers a holistic perspective that connects being, doing, and knowing, and provides a good account of how knowledge, agency, and cultural processes are intertwined. This approach has been gaining ground in recent decades, is naturalistic but not reductionist, and draws on sources from phenomenology, biology, complexity sciences, and cognitive sciences (Varela et al. 2017; Thomson 2007; Stewart et al. 2010; Di Paolo 2021; Di Paolo and De Jaegher 2012, 2022; Di Paolo et al. 2023; Sepúlveda-Pedro 2023). It distinguishes itself from older approaches that view cognition as informational processing based on internal representations. In contrast, the enactive approach sees cognition as an emergent process in the interaction between the organism and the environment, giving greater prominence to the whole body, in which the mind is embodied in perception and thought. Anticipation, viewed from this vantage point, is deeply embedded in action, in lived experience. Living beings react to the world around them and anticipate it in ways that allow them to adapt to and navigate their environment proactively. This anticipation operates both at the cognitive and physical levels through sensorimotor coupling. Organisms anticipate as they actively interact with the environment through patterns of behavior that have proven effective in the past. Anticipatory action generates consequences, which in turn inform and shape future actions.

The enactive approach recognizes the precarious nature of these processes, which are invariably prone to error, a certain amount of uncertainty, and emergent phenomena. Anticipation does not necessarily produce a predetermined action or expected outcome; instead, it guides the organism in its ongoing interaction with the world. From this perspective, anticipation is rooted in action dynamics (*praxis*). It is part of a continuous process of adjustment and adaptation involving the whole organism with its environment, as opposed to isolated brain processes. Here, mind and body are intertwined, and the whole organism is intertwined with the world in a co-creation process. Enactive cognition suggests that living organisms actively forge and maintain their identities through interacting with the world around them. Our way of knowing the world is based on something other than anticipating it, as the theories of consciousness and emotion mentioned above suggest. Instead, we are co-creators of the world, "bringing forth" only what is relevant to us.

Thus, doing and knowing are recognized as ways for organisms to "be" alive. Organisms "bring into being" or "make real" their world through their interactions with it. Perception is not merely a passive representation of reality; we perceive the world through our potential actions. We perceive things not only according to their attributes but also by what we can do with them.

A crucial element of this approach is the link between anticipation and "sense-making," whereby the organism interprets its environment in the context of its "affordances." These affordances arise from the interplay between the environment and the organism's capabilities, influenced by biological requirements and the accumulated experience of interactions. From an affective or valuational point of view, this perspective holds that the world has no intrinsic meaning independent of organisms. Organisms are embodied agents that create meaning (*enact*) through these interactions with the world. What is vital to one being may be of no consequence to another. Sense gives meaning and relevance, guides behavior, and creates a set of norms. Therefore, action is intentional and directed toward perpetuating one's life and realizing one's goals by what happens in the surrounding environment.

Enaction emphasizes continuity and innovation among natural processes, life, the mind, language, and human communities. For human beings, agency manifests itself in deeds, intentions, emotions, thoughts, relationships with others, speech, community involvement, and work. *"Embodied agents* enact *a world of significance, an ongoing achievement done in conjunction with other agents. Acts, moreover, are processual; they respond to existing potentialities, trigger or block processes of individuation, and have rippling repercussions in a meshwork of other acts past and future"* (Di Paulo and De Jaegher 2022).

Human beings are constantly formed, whose identity and understanding of the world are forged through an interactive process of attributing meaning, or "sense-making." It is not a purely individual process but one deeply rooted in social and cultural dynamics, where emotions are interwoven and shared and can, in turn, be in harmony or conflict, with unequal effects on our life chances. Through language, our actions are transformed into forms of knowledge and power, encompassing practical skills, ethical reasoning, and making and breaking habits and norms. These human actions are an integral part of the values that are the basis for the success or failure of our social interactions. They are also connected to a more extensive network that includes the subtleties of behavior, such as attitude, style, or mood. In this way, they form the ethical reality that must be continually updated.

This framework can accommodate phenomena such as hope or commitment and complex and profound states beyond the conscious anticipation of future events. In this light, hope is not a definite expectation as it is a guiding sense toward something inherently good that cannot be fully understood or anticipated. We are inspired by those who exhibit hope or courage, awakening previously hidden potentialities and influencing our actions from a pre-rational stage. Hope and commitment can transform what might be called "pre-individual potentialities," charging them with an energy that has not yet manifested itself but is already directing our moral compass and, thereby, our actions. Similarly, acts of participation bring about a shift in collective consciousness, which involves not only a cognitive shift but also a change in ethical possibilities, giving life to new practices and thus to new realities (Di Paulo and De Jaegher 2022).

Here is where the concept of "hopeful determination" makes sense as a way of being doing knowing that implies an active commitment to shaping a desired future. It transcends the individual and weaves its way into webs of interconnectedness and affect. As Motti (2017) notes, this resonates deeply with the rhythms of Deleuze (1994) and involves the co-creation of meaning with others and the world in a context of uncertainty. "Hopeful determination" perseveres in the face of adversity through adaptability and resilience in an ongoing effort to transform potentialities into concrete realities (Coleman 2017; Cook and Cuervo 2019; Doe 2020; Benton 2019). It emerges as a vital component in all human endeavors, cities being among the most complex, playing a decisive role in their current and future configuration.

3.3 Hopeful Determination in the Urban Context

Cities, seen through the lens of "things," emerge as hyper-complex systems, a mesh of deterministic inertias and emergent phenomena. To the outside observer, they represent a confluence of chance and necessity (Prigogine and Stengers 1979), whose trajectory could be more or less reliably anticipated using various methods (Miller et al. 2013). That would depend on the magnitude of this inertia (demographic patterns, economic and technological infrastructure, communications, institutions, social and cultural habits) and the stability of their operating context (the surrounding political, social, and environmental trends).

Looking at the city from an action perspective, however, adds a crucial element to the interplay between chance and necessity: the intricate dynamics of the conscious will of human actors and their efforts to steer the future (Godet and Durance 1997) toward a desired destination, guided by their values and expectations, or what we refer to here as their "hopeful determination." That will is crucial in urban planning, as it stems from prevailing visions of the future. These visions, in turn, help to mitigate uncertainty by enabling the actions of urban actors to be coordinated, fostering stability and predictability, and allowing for a certain level of foresight in the short and medium term. Thus, the co-creation of the future in urban environments emerges from the interplay between the capabilities (knowledge and power) of the different actors and groups involved in urban governance, all striving to realize these visions of the future.

A particularly heuristic model in this regard is the three-horizon model (Sharpe 2013; Sharpe et al. 2016), which aims to identify the underlying patterns in actors' behavior, i.e., their value systems and attitudes toward continuity and change, or, in the author's words, their "patterns of hope."

The three-horizon model shines a bright light on how three essential visions unfold at every unique intersection of actors in space and time. Horizon 1 reflects the perspectives of those actors who, clinging to control and management of the current system, resist change. Nevertheless, they are propelled into a future that may be unsustainable, undesirable, or both, and face the dichotomy of radically transforming themselves or fading into oblivion. Horizon 3, on the other hand, represents the visions of the more farsighted agents of change, those who are inclined to create favorable scenarios based on principles of fairness and environmental sustainability but who are hampered by a lack of resources to make a substantial impact in the short, medium, or long term. Finally, Horizon 2 embodies the visions of those pragmatic enough to combine Horizon 1 resources and Horizon 3 aspirations to effect significant change toward a preferred course. However, they risk being overwhelmed by the inertia of those in Horizon 1, who may allow change only if they maintain their dominant position.

In periods of relative stability, when urban socio-technical systems and their environment are fully embedded but unsustainable in the long term, challengers to the system face more significant obstacles. However, in internal or external turbulence threatening to destabilize existing relationships, Horizon 3 and 2 players can see a window for change. Such moments form the basis of a multi-level perspective on change and innovation that clarifies how transition occurs across different levels: the global landscape or context, socio-technical systems, and innovation niches (Geels 2002, 2007, 2010, 2011; Geels and Schot 2007).

In transformative processes of this kind, over and above the availability of material resources, narratives (Liveley 2017) play a critical role in inspiring and galvanizing people, giving meaning to individual and collective actions, and injecting "hopeful determination." As Donella Meadows (2008) observes, the most effective lever for changing a system is changing the underlying paradigm, the meta-narrative that frames the stories of the various actors. Examples of urban narratives include the Creative City, the Business City, the 15-Minute City, the Mediterranean City, and the City of Care. Such narratives spring from a profoundly uneven field, where perhaps the most telling indicator of the power of urban actors is their ability to make their vision and narratives prevail, whether through persuasion or imposition (González-Arellano 2022). Horizon 1 elites, with operations and resources that extend beyond city boundaries on many levels, often wield tremendous influence over information and communication systems. These systems play a pivotal role in generating and perpetuating meaning and making sense of life and the individual and collective values that, in turn, influence urban behaviors and decisions. They are, therefore, in a position to influence who and what "will have a future" in the city, what will remain unchanged, and what needs to evolve. These patterns uniquely shape each urban context, reflecting the complexity of particular circumstances and multi-actor, multi-level processes.

Therefore, the "narratives of hope" map an uneven future vision across the urban fabric. In underprivileged neighborhoods, not only are current conditions less than ideal but there is also the sense of the future being at a dead end. These environments, aggravated by heightened tensions and acute instability, offer remarkable opportunities for bringing about change and consolidating processes of transformative "hopeful determination." The uniqueness of each context adds value to case studies such as the one presented below. With them, we can unravel how horizons manifest and converge within distinct landscapes while also uncovering insights that help to identify environments and practices capable of expressing the will of different actors in different contexts. The result is a climate of inspiration and unwavering hope that catalyzes the activation of their skills.

3.4 Patterns of Hope on the Ground: The Integrated Plan for Huelva's District 5

The Integrated Plan for District 5 of the City of Huelva (PID5H) is a paradigmatic case of urban foresight planning worthy of study, both for its complexity in terms of the issues addressed and its participatory process, as well as the fact that it has lasted for more than two decades. This case provides an opportunity to analyze, in an evolutionary context, the interplay of "hopeful determinations" of different actors (their active commitment to co-create a desired future). It illustrates how these are translated into collaborative courses of action of varying intensity, depending on contextual factors, individual circumstances, visions, and collective sense-making processes organized around an overarching goal. It also makes it possible to identify the main competencies acquired by the actors in their actions, developing their individual and collective agency in an active process in which being, doing, and knowing are inextricably linked.

3.4.1 Origin and Characteristics of the PID5H

Huelva (population 141,824) is located in southwest Spain. The PID5H[1] was established in 2000 as a comprehensive urban renewal plan in the fight against poverty and social exclusion. It aims to eradicate discrimination and safeguard the dignity, quality of life, and autonomy of all the people living in seven neighborhoods (approx. 15,000 inhabitants). The area was characterized by extreme insecurity and vulnerability, having been neglected by the public authorities, but there was a certain, albeit fragmented and uncoordinated, network of social initiative organizations.

The initiative was conceived by social organizations working in the area in close collaboration with a research center at the University of Huelva. Together, they crafted a narrative about the urgent need to strengthen and coordinate efforts to bring about social change in this neighborhood. In doing so, they garnered the support of local neighborhood associations and persuaded the public authorities to get involved and allocate resources. The primary vehicle used was a multi-actor platform involving public bodies from education, health, housing, employment, social services, culture and social initiatives, businesses, neighborhood associations, state-subsidized schools, and sports clubs. This multi-level platform—national, regional, and local—involved more than 50 entities (now 73). Its participatory governance system is still an internationally recognized benchmark for partnership building, holistic visioning, multi-sectoral coordination, and citizen participation in the regeneration of neighborhoods in need of social transformation (Miedes-Ugarte and Fernández-Borrero 2010; De Paz et al. 2017).

As the very name implies, the methodological pillars of PID5H are its multidimensionality (holistic approach), multi-actor and multi-level partnerships, citizen participation, and long-term visioning. Its goals, lines of action, and *modus operandi* have been set out in various strategic plans (2000, 2008, 2016, 2023), each time informed by preparatory anticipatory work carried out collaboratively. This work combines a thorough diagnosis of the current situation with a forward-looking analysis of expected trends and desired scenarios in the chosen time horizon.

The governance model is similar to a *holocratic* organization's (Robertson 2007), where decision-making power is pooled and shared. Activities are organized into six thematic participation forums (citizenship and planet; employment and prosperity; health and healthy habits; education and training; children and youth; family, equality and integration). These autonomous, self-organizing work units create ad hoc commissions for specific projects as needs or opportunities arise and for as long as they last. Each forum, which meets quarterly, defines the role of the coordinator, who in turn liaises with other forum coordinators and representatives of regional and local public authorities on the Steering Committee at monthly follow-up meetings.

Decisions in the forums are made by consensus. Discussions continue until resolutions are ready to be proposed to the Steering Committee, where a new consensus must be reached. Decision-making power is decentralized throughout the organization and is not in the hands of a single person or group. All the organizations

1 All the information regarding the PID5H in this chapter can be found at www.distrito5huelva.org.

and individuals involved in the forums come together at the annual meeting of the General Assembly, the supreme representative and participatory body, which ratifies the membership of the various partners. Information is shared openly and readily available to all organization members. That is a very dynamic model of governance that has been slightly adapted over the years as circumstances dictated.

The Plan's entire operation is orchestrated by a technical office with two facilitators responsible for coordinating, communicating, planning, and sustaining the participatory process of collective responsibility. In addition, a technical team led by two members of the University (the authors, since 2010) provides scientific and technical advice throughout the process. The Plan has no legal status, and aside from a few small joint projects, the technical office's sole function is to coordinate the visions and actions of the various entities. Each entity develops its projects and manages its funds: public grants and private contributions from competitive tendering fund the technical office.

3.4.2 Phases of PID5H: Patterns of Hope

Over the past two decades, the PID5H has weathered many ups and downs, and it is possible to distinguish between different phases in its evolution based on several criteria. These include the changing needs of a rapidly aging population and new challenges such as climate change; leadership changes and vacuums within and between entities; changes in the people involved and in the very lives of the long-serving participants; changes in funding bodies; shifts in public policy approaches; and the evolving landscape of policy direction at the national, regional, and local government levels, with the attendant agreements and conflicts. Of particular value to this analysis is the aforementioned three-horizon model for its ability to heuristically identify the meta-narratives underpinning the ways how different actors engage with the PID5H, or in the author's own words, "their patterns of hope" (Sharpe 2013). How the meaning and significance of the PID5H has evolved for the various actors goes a long way toward explaining the progress and setbacks observed throughout the collaborative process.

Initial Phase (2000–2009): Horizon 3 with Support from Horizon 2

A typical pattern emerges among all the actors involved in this initial phase, clearly recognizable as Horizon 2. This pattern serves as a major impetus for innovation for Horizon 3 actors, who have been striving for years to create a better future in living conditions for the population but with little support and scarce resources. In this phase, significant changes are negotiated and approved, including establishing the participatory process and large-scale projects such as constructing a state-of-the-art health center or rehabilitating and bringing up to code part of the housing stock. The visions of the actors (government, neighborhood associations, and other entities) are strongly aligned during this phase. Naturally, conflicts of varying intensity and nature exist, especially as the entities get to know each other and learn to work together. The prevailing mood, however, can be described as one of shared "hopeful determination." The ultimate proof is that despite the challenges, the organizations remain in the game (Kahane 2017), with all renewing their commitment to participate in the Plan every year and more and more players coming on board.

Second Phase (2010–2020): Horizon 1 Reels in Horizon 2

The financial crisis in Spain worsened from 2010 onwards, and this has had a significant impact on all PID5H actors. Social organizations are left reeling by the crisis, turning all their attention inward to weather the storm. Some of them disappear. The public sector, mainly social services, has fallen victim to drastic funding cutbacks. In a context of great uncertainty, the gravity of the financial crisis has prompted a return to more welfare-oriented policies to combat poverty (Horizon 1) at the expense of the PID5H's more proactive and anticipatory policies (Horizon 2). Neighborhood associations focus on strengthening their fragile networks of community solidarity to respond to the worst neighbors' situations. They see the PID5H as an indispensable resource for further empowerment, reinforcing these networks and stepping up efforts when needed. Under these circumstances, public authorities, whose leaders are overwhelmed by the circumstances and in constant flux either need help to understand the operating logic of the PID5H or perceive the participatory process to be a threat due to increasing neighborhood demands. The neighborhood associations, with their increasingly aging representatives, point out the shortcomings and the lack of response to issues about which, after ten years of constant attention and collaboration, they have created a collective memory, have expert knowledge, and have grown weary of calling for concrete solutions. Partnership working, when it occurs, continues to bear fruit, but tensions are rising.

That is compounded by financial difficulties in maintaining the technical office, which has gone months without funding for the first time in a decade. The sustainability of the mechanism itself becomes a significant concern for some Plan members. Others, faced with existential threats of their own, do not see it as a priority. All of this weakens the participatory process, causing the Horizon 2 narrative to recede and the process itself to retreat to a more management-oriented Horizon 1 in an attempt to keep the process alive under adverse conditions.

Third Phase (2020–2023): The Return of Horizon 3 (or the Jedi)

Despite the progress and setbacks experienced by the actors in the area, there has always been a latent critical mass that can be identified with Horizon 3. Their determination sees greater potential in the environment for transitioning to a better future. In a paradoxical twist, the COVID-19 pandemic brings these actors closer together, prompting a joint reflection on the need to focus on the local population's reality and to renew the Plan's long-term, forward-looking vision. New challenges, including but not limited to energy transition, climate change, population aging, and technological adaptation, join the list of existing problems to be solved. This renewal breathes new life into the transformation narrative and is deeply meaningful to the various actors, rallying visions, resolve, and resources around the common goal of reigniting the transformative innovation of Horizon 2. Building on this momentum, a new participatory foresight and adaptive planning process called Roadmap 2030 seeks to recapture the most valuable lessons of the past two decades. The process has involved the creation of five citizen laboratories inspired by the aforementioned three-horizon approach. Combined with a mission-oriented innovation methodology (Mazzucato 2021), this has given concrete form to the future visions of the actors by identifying 37 projects organized into six major missions (one for each forum). The aim is to mobilize the attention and energy of the alliance around meaningful and hopeful common goals.

This new narrative, co-constructed by more than 200 people and endorsed by the 2022 Assembly, will reinvigorate the participatory process. The first step will be to find new funding sources to sustain it while introducing fresh approaches to how actors share enabling roles and forging new alliances between organizations, neighbors, and public bodies following a period marred by misunderstanding. To this end, the projects have also been aligned with the 17 sustainable development goals of the UN 2030 Agenda and the ten goals of the Spanish Urban Agenda, defined by the local government for the city as a whole (Miedes-Ugarte and Fernández-Borrero 2022).

The complexity of the PID5H's evolution through these phases shows that the journey to the desired future is non-linear, with the tapestry of visions reconfiguring along the way. This journey spans the period of initial alignment and innovation, where Horizon 2 drove development and participation, through the struggle for survival and the resurgence of welfare policies during the financial crisis (Horizon 1), to the period of reflection and the search for a new, forward-looking direction in the post-pandemic era (Horizon 3). All of this profoundly influenced the development of the actors' competencies, which have co-evolved throughout the process in terms of their anticipatory capacity and ability to work in partnership.

3.4.3 The Competencies Developed by the Actors Through an Enactive Lens

In each of these three phases, opportunities have arisen for actors to learn how to enhance their foresight skills and, by extension, their ability to meet future challenges. In the initial phase (2000–2009), Horizon 2 provided a transformative impulse that strengthened the actors' anticipation by constructing major infrastructure and high-community impact projects. This period was critical in laying a foundation of cooperation and trust that, by removing some of the uncertainty, proved essential to improving individual and collective anticipation. Despite the tensions, aligning actors' visions helped develop a proactive approach to identifying and addressing future problems.

In the second phase (2010–2020), the financial crisis forced actors to prioritize short-term survival over innovation. That diminished their anticipatory capacity for an extended period, as resources and attention were focused on immediate needs rather than long-term planning. Nevertheless, dealing with the crisis also provided valuable lessons in risk management and the need for flexibility in existing plans and structures.

The third phase (2020–2023), marked by the pandemic and a renewed focus on Horizon 3, was an opportunity to revisit and enhance anticipatory capacity. The now more seasoned actors have successfully incorporated emerging issues such as climate change and energy transition, reflecting their evolving ability to anticipate and build a shared vision of the future.

Developing these anticipatory skills is part of a broader set of competencies needed for the ongoing practice of "hopeful determination," learning to work together to meet future challenges with flexibility and adaptability, depending on goals and circumstances (Castañer and Oliveira 2020). Four main groups of competencies have been identified:

1. Proactive collaboration: alliance building, conflict management, effective communication.
2. Adaptation and change management: adaptability and flexibility, managing uncertainty, organizational and community resilience.
3. Innovation and transformation: innovation: systems thinking, creative thinking, adaptive planning, resource management, and fundraising.
4. Social engagement and community empowerment: civic engagement, socio-ecological awareness and responsibility.

Developing these competencies is by no means technically or formally linear. It is a complex, open-ended, never-ending process that occurs under precarious conditions and is unevenly distributed. It is a process that strongly resonates with the enactive approach to cognition. This approach views cognition not simply as a representation of the world but as a collection of embodied activities that emerge and are continually transformed through interaction with the environment. The path taken by PID5H is a real-life example of the enactive principle, where cognition and action are in constant dialogue, and every action is both a product and an enabler of understanding. The actors' acquired competencies can be seen as manifestations of their unfinished agency in continuous co-construction with the environment. In an ongoing process of sense-making, actors shape the environment and are shaped by it. Far from merely receiving or processing external information, they have developed the ability to read subtle signals (Morales and González 2021), adapt to emerging circumstances, and proactively seek to influence events in a positive direction. Anticipation here means actively participating in creating desirable futures, not merely predicting what is to come.

3.5 Conclusions

The future is in seed form right now; the will of urban actors is guided by their focus, drawn to the possibilities offered by an environment aligned with individual and collective sense-making in specific spatial and temporal contexts. We have argued that "hopeful determination" catalyzes the actors' impulse to channel their energies toward a more promising future within this sense-making framework. The enactive approach emphasizes that the future is flexible. Instead, it is an emergent entity forged by our present actions and not entirely predestined. Actors and their environment co-create worlds of meaning. These worlds steer the actors' activity toward realizing their aspirations. They transform the possibilities within the environment into tangible realities.

The analysis of PID5H illustrates a form of active and participatory anticipation that, according to Gidley's (2010, 2016) categorization, taken up by Motti (2017), falls into the "empowerment-participatory and hope-enhancing" category. It is a living process, full of uncertainties, in which all possible combinations of challenges involved in transforming neighborhoods in crisis can be found in varying degrees of intensity: the scarcity of resources, stigmatization of specific areas, limited acknowledgment and visibility of the process by the actors, uneven distribution of capacities, fragile citizen participation in continual development, resistance of some actors to change, and varying institutional support depending on whether participatory solutions conform to pre-established policies, not to mention bureaucratic barriers. It takes great determination and hope to continue working together in such a context. The PID5H exemplifies inclusive and caring governance that seeks to sustain this "hopeful determination." At the same time, it considers the emotional and affective dimensions of the participants and the community, who often face specific and ongoing challenges that affect their minds and ability to act. In a world grappling with complex socio-ecological challenges like climate change, political polarization, and post-truth, and where advances like genetic engineering and artificial intelligence are unleashing massive transformative potential at great risk of deepening inequalities, the enactive approach reminds us of the importance of reconnecting with our embodiment and with the unique textures of our interactions. That makes the sensory experience of participatory spaces highly relevant. It triggers potential connections between aspects essential for collaborative and anticipatory behavior, such as memory, foresight, voice, optimism, and yearning (Bussey 2017).

This chapter also highlights how actors construct meaning through the mobilization of narratives. As Mercier and Sperber (2017) suggest, the essence of sense-making comes from our interactions and conversations, which are the underlying catalysts for transformation. In this sense, as Kaplan and Davidoff (2014) note in their exploration of the concept of "delicate activism," the main transformative lever in many situations is to change the nature of the conversation. That is consistent with O'Hara and Leicester's (2019) proposition that the three literacies—*cognitive* (understanding how we know), *psychological* (understanding our agency), and *cultural* (understanding and facilitating cultural change)—should be strengthened as essential competencies for navigating a VUCA world with some guarantee of success. As we have shown, the enactive approach can significantly contribute to this reinforcement.

In short, anticipation in the urban context discussed here does not negate the value of analyzing expected trends or scenarios in more or less stable contexts, especially for risk prevention purposes. However, its effectiveness is inextricably linked to the actors' willingness to act accordingly and adapt their behavior. They must also activate their individual and collective agency, an ongoing process, to positively influence the future and break with the inertia and trends of the past when they are unsustainable or undesirable. It is about seeing the future as a shared legacy constantly co-constructed in an ongoing process of renewing "hopeful determination" or, as Solnit (2010) would say, as a never-ending effort to maintain and sharpen the axe with which to open emergency doors.

References

Barrett LF (2017) How emotions are made: the secret life of the brain. Pan Macmillan

Benton M (2019) Epistemological aspects of hope. In: Blöser C, Stahl T (eds) The moral psychology of hope. Rowman & Littlefield Publishers. Rowman and Littlefield International, pp 135–151

Bussey M (2017) Anticipatory aesthetics: the senses and the body in anticipatory theory and practice. In: Poli R (ed) Handbook of anticipation: theoretical and applied aspects of the use of future in decision making. Springer, pp 861–874

Castañer X, Oliveira N (2020) Collaboration, coordination, and cooperation among organizations: establishing the distinctive meanings of these terms through a systematic literature review. J Manag 46(6):965–1001. https://doi.org/10.1177/0149206320901565

Coleman R (2017) A sensory sociology of the future: affect, hope, and inventive methodologies. Sociol Rev 65(3):525–543

Cook J, Cuervo H (2019) Agency, futurity, and representation: conceptualising hope in recent sociological work. Sociol Rev 67(5):1102–1117

Deleuze G (1994) Difference and repetition. Columbia University Press

De Paz MA, Asensio MJ, Franco MI (2017) Plan integral Distrito V de Huelva: 2000/2003, vol. 87Servicio de Publicaciones de la Universidad de Huelva

Di Paolo EA (2021) Enactive becoming. Phenomenol Cogn Sci 20(5):783–809

Di Paolo E, De Jaegher H (2012) The interactive brain hypothesis. Front Hum Neurosci 6:163

Di Paolo EA, De Jaegher H (2022) Enactive ethics: difference becoming participation. Topoi 41(2):241–256

Di Paolo EA, Lawler D, Vaccari AP (2023) Toward an enactive conception of productive practices: beyond material agency. Philos Technol 36(2):31

Doe MJ (2020) Conceptual foreknowings: an integrative review of hope. Nurs Sci Q 33(1):55–64

Geels FW (2002) Technological transitions as evolutionary reconfiguration processes: a multi-level perspective and a case-study. Res Policy 31(8–9):1257–1274

Geels FW (2007) Feelings of discontent and the promise of middle range theory for STS: examples from technology dynamics. Sci Technol Human Values 2(6):627–651

Geels FW (2010) Ontologies, socio-technical transitions (to sustainability), and the multi-level perspective. Res Policy 39(4):495–510

Geels FW (2011) The multi-level perspective on sustainability transitions: responses to seven criticisms. Environ Innov Soc Trans 1(1):24–40

Geels FW, Schot J (2007) Typology of sociotechnical transition pathways. Res Policy 36(3):399–417

Gidley JM (2010) Globally scanning for 'Megatrends of the Mind': potential futures of futures thinking. Futures 42(10):1040–1048
Gidley JM (2016) Postformal education: a philosophy for complex futures. Springer International Publishing, New York, NY
Godet M, Durance P (1997) La prospective stratégique. Futuribles 45:57
González-Arellano S (2022) The future scenarios of cities: an analysis of their institutional construction. Int J Urban Reg Res. https://doi.org/10.1111/1468-2427.13103
Kahane A (2017) Collaborating with the Enemy. Berrett-Koehler Publishers
Kaplan A, Davidoff S (2014) A delicate activism: a radical approach to change. The proteus iniciative. http://socialself.org/Articles/adelicate.aspx
Liveley G (2017) Anticipation and narratology. In: Poli R (ed) Handbook of anticipation: Theoretical and applied aspects of the use of future in decision making. Springer, pp 1–120
Mazzucato M (2021) Mission economy: A moonshot guide to changing capitalism. Penguin UK
Meadows DH (2008) Thinking in systems: a primer. Chelsea Green Publishing
Mercier H, Sperber D (2017) The enigma of reason. Harvard University Press
Miedes-Ugarte B, Fernández-Borrero MA (2010) Inteligencia Territorial para la Lucha Contra la Pobreza. Aprendizajes de 20 Años Sobre el Terreno. TRABAJO. Revista Iberoamericana de Relaciones Laborales, p 23
Miedes-Ugarte B, Fernández-Borrero MA (2022) IV Plan Integral Distrito 5. Hola de Ruta 2030. from https://www.distrito5huelva.org/iv-plan-integral-del-distrito-5/
Miller R, Poli R, Rossel P (2013) The discipline of anticipation: exploring key issues. Bellagio Document 4 Working paper 1 version 0.9
Morales N, González S (2021) Cities as anticipatory systems: analyzing 'weak signals' to explore beyond the predictability of their future. Ethnographic Praxis Ind Conf Proc 2021(1):240–257
Motti VV (2017) Sources of futures studies from foresight to anticipation. Poli R (ed) Handbook of anticipation: theoretical and applied aspects of the use of future in decision-making. Springer, pp 1719–1732
O'Hara M, Leicester G (2012) Dancing at the edge: competence, culture, and organization in the 21st century. Triarchy Press
O'Hara M, Leicester G (2019) Dancing at the edge: competence, culture, and organization in the 21st century. Triarchy Press
Peterson JB (2002) Maps of meaning: the architecture of belief. Routledge
Poli R (ed) (2017) Handbook of anticipation: theoretical and applied aspects of the use of future in decision making. Springer
Prigogine I, Stengers I (1979) La nouvelle alliance: métamorphose de la science. Gallimard
Robertson BJ (2007) Organization at the leading edge: introducing Holacracy™. Integr Leadersh Rev 7(3):1–13
Sepúlveda-Pedro MA (2023) Sense-Making in the wild: the historical and ecological depth of enactive processes of life and cognition. Adaptive Behavior, 10597123231190153
Seth A (2021) Being you: a new science of consciousness. Penguin
Sharpe B (2013) Three horizons. Triarchy Press
Sharpe B, Hodgson A, Leicester G, Lyon A, Fazey I (2016) Three horizons: a pathways practice for transformation. Ecol Soc 21(2)
Solnit R (2010) Hope in the dark: the untold history of people power. Canongate Books
Stewart J, Gapenne O, Di Paolo EA (eds) (2010) Enaction: toward a new paradigm for cognitive science. MIT Press
Thompson E (2007) Mind in life: biology, phenomenology, and the sciences of mind. Harvard University Press
Varela FJ, Thompson E, Rosch E (2017) The embodied mind, revised. MIT Press, Cognitive science and human experience

Part III
New Building Anticipatory Systems in Urban Sectors in the Global South

Chapter 4
Sustainable Urban Mobility as an Anticipatory System: The Case of the Quito Metro

Marco Córdova and Diana Paz

Abstract Sustainable urban mobility is presented as an anticipatory system in urban development processes. The future projection of socially inclusive, environmentally sustainable, and economically efficient territorial configurations allows for the improvement of public policies and city planning in the present. The chapter proposes an analysis of how the Quito Metro constitutes an anticipatory system in the growth of the metropolitan district, in the mitigation and adaptation to climate change, and in the socio-cultural dynamics of the population. Methodologically, a descriptive inference exercise is proposed through a qualitative case study. The collection and processing of information are based on academic secondary sources and official documents. The case of Quito shows that the Metro is built with an anticipatory approach, arising from a more comprehensive reading of mobility as a long-term policy. However, it constitutes an implicit or prospective anticipation based on the city's planning model.

Keywords Sustainable development · Urban mobility · Metro systems · Anticipatory systems · Metropolitan district of Quito

4.1 Introduction

In recent decades, socio-ecological transitions have driven territorial strategies based on sustainable development approaches. These strategies have redefined issues such as climate change, risk due to natural hazards, natural resource management, and urban mobility. Faced with the imperative to build sustainable and resilient territories, multi-stakeholder (state, market, society) and multi-level (international, regional, local) governance dynamics have been consolidated. In this context, various state

M. Córdova (✉) · D. Paz
FLACSO Ecuador, Quito, Ecuador
e-mail: mcordova@flacso.edu.ec

D. Paz
e-mail: dmpazfl@flacso.edu.ec

S. Gonzalez-Arellano and B. Gandlgruber (eds.), *Cities as Anticipatory Systems*, The Urban Book Series, https://doi.org/10.1007/978-3-031-78162-9_4

and non-state actors have promoted a comprehensive set of public policies and planning instruments. These new forms of public action can be understood as anticipatory systems—sets of decisions made in the present based on future resources and scenarios (Poli 2017).

The contemporary notion of sustainable development has a constructivist and holistic sense. It is projected into the future regarding social equity, environmental protection, and economic efficiency, with a marked impact on territorial configuration and transformation (Gudynas 2010; Keivani 2010). Viewing development ontologically from an anticipatory discipline perspective allows understanding the planning exercise derived from sustainable development as a form of knowledge aimed at colonizing the future and giving meaning to the emerging novelty of the present by inspiring a systematically imagined future (Miller et al. 2013). This anticipatory reading of development in a sociological key implies understanding that knowledge about the future addresses the diagnosis of the trends of the times and simultaneously projects, in a speculative sense, possible scenarios and outcomes of change (Hammershøj 2017).

Sustainable development has been built on the positive and negative interpretation of the territorial transformations of the last century, explaining the trends of the transitions derived from the process, especially in contexts where structural imbalances are more evident (Acosta 2015; Bolay 2012; Winchester 2006). However, at the same time, the idea of development based on sustainability principles has led to the institutional construction of a normative, regulatory, and cultural-cognitive nature, which, through the agenda of the Sustainable Development Goals (SDGs), condenses a set of anticipatory policies that local governments have promoted to model future scenarios (González-Arellano 2022).

Analyzing the processes of public action inherent to sustainable development as a problem of anticipation allows us to understand how the utopian construction of the territories of the future is being employed in the present to enhance the design of public policies in the face of urban problems in the current context of the Anthropocene. In the field of urban mobility, from the positioning of sustainable development in the 1990s, a change is observed from previous transportation policies oriented in an engineering approach toward a new mobility paradigm that includes the relationship of the individual with space (Miralles-Guasch and Cebollada 2009). This integration addresses problems related to the unequal development of cities, later conceived from economic, social, and environmental sustainability (Quintero-González 2017; Vasconcellos 2015).

Around sustainable urban mobility in Latin America and the Caribbean (LAC), a series of policies have been promoted, focusing on accessibility, equity, integrality, and emissions reduction (Errazuriz et al. 2017; Figueroa 2001; Rivas et al. 2019). More recently, attention has turned to the link between urban development and mobility in the so-called Transport-Oriented Development (Rodriguez and Vergel-Tovar 2018). In these processes, implementing Metro systems has promoted socio-technological innovations in public transportation and urban mobility in general, serving as a central element in new ways of conceiving the city. In this sense, mobility policies are being consolidated as anticipatory systems in LAC cities based on their intention to manage uncertainty and foster the collective intelligence of cities (González-Arellano 2022). The notion of forward-looking sustainability allows for structuring comprehensive urban development models in the present, characterized by complex contexts with scarce resources, low institutionalization, and weak capacities (Paz-Gómez and Santelices 2020).

Against this background, the chapter analyzes how sustainable mobility, specifically Metro systems, configures anticipatory systems in urban development. The Metropolitan District of Quito (DMQ) is selected as a case study since it has been promoting a sustainable mobility policy that has transformed urban planning since the 1990s. Implementing the first Metro line in 2023 marks a shift in the trajectory of mobility policies, primarily redefining current strategies based on the future projection of the city's development. Methodologically, the proposed combines a contextual analysis (LAC) and a case study (DMQ) to achieve a descriptive inference level and take a qualitative approach. The collection and systematization of information are based on secondary sources related to academic publications, official documents, and reports on development and land use plans, mobility policies, and impact analysis, among others.

The chapter is structured in two parts: (i) the characterization of urban mobility and Metro systems in LAC, based on a review of the specialized literature of the region, and (ii) the analysis of the case of the DMQ in terms of the background of the incorporation of the Quito Metro as a public transportation hub, and its impact as an anticipatory system in urban development, in the mitigation and adaptation to climate change, and the socio-cultural dynamics of the city.

4.2 Sustainable Urban Mobility in LAC: From Reaction to Anticipation

Urban mobility has been associated with cities' growth and displacement dynamics. Data suggest that 80% of the population lives in cities, with LAC concentrating on highly dense cities such as Mexico City, Buenos Aires, Rio de Janeiro, Sao Paulo, Bogota, Lima, and Santiago de Chile (CEPAL 2018). The region maintained economic growth between 1995 and 2017, which incentivized the increase of the private transport fleet (CEPAL 2019). The discussion on mobility has been conceived from two main orientations. The first is related to the reaction to the urban system's road, transport, or energy dynamics, and the second involves the reconfiguration of mobility toward anticipation, derived from adopting the sustainability paradigm in urban design and planning.

Urban mobility, defined as a reactive system, was oriented by policies focused on responding to the growth of cities. This model generally supports an engineering vision that responds to the dynamics of "the greater the growth, the greater the road development." It is characterized by short-term governmental responses designed to solve the problems derived from rural–urban migration, new patterns of land occupation, and travel times in origin–destination trips (Casado Izquierdo 2008). This reactive mobility system prioritized analyses on intra-urban commuting modes (Cosacov and Di Virgilio 2018). Although traffic engineering incorporated studies on supply and demand, origin–destination surveys, vehicular flows, and transport modes to forecast future demands, mobility was not considered a response to a possible evolution. Instead, it addressed the need to move within the city with infrastructure and transport (Manheim 1979; Quintero 2011; Rodríguez Vignoli 2008).

This generation of engineering resulted in the construction of new roads and ring roads, leading to increased private transportation. Although there was an occasional increase in public transport services, these were characterized by private operations with little or no state regulation (Avellaneda 2017; CAF 2011). Within the framework of reactive responses, public transport was constituted as a tool to supply the demand for transportation in line with population and economic growth (Lupano and Sánchez 2009). Thus, in the 1990s, the boom of rapid transit buses and omnibuses did not incorporate an environmental sustainability approach. Once again, the short-term vision of implementing new transportation systems was associated with supply and demand and, to a lesser extent, with prospective planning considering the impacts of population growth and urban expansion.

This reactive urban mobility underwent its first change with the anthropocentric incorporation of displacements. In this sense, mobility focused on relating social and cultural dynamics to how individuals were transported (Levy 1999; Sheller and Urry 2006). That did not imply leaving aside the analysis tools already designed by transportation engineering but incorporated analyses on supply and service conditions (Gutiérrez 2012; Miralles-Guash 2002; Montezuma 2003), territorial, sociodemographic, and residential conditions, and travel frequencies and their distances (Cebollada and Avellaneda 2008; Lazo 2008). Furthermore, daily mobility integrates variables associated with providing goods and services, considering transportation as the planning axis. Subsequently, these analyses incorporated variables such as the environment aimed at understanding the impact of transportation use on pollution levels in cities, accessibility and proximity to goods and services, and their impact on travel costs, residential preferences, and territorial planning (Estevan and Sanz 1996; Hernández and Falavigna 2016).

The second conception of mobility is associated with incorporating the sustainability paradigm, which consolidates an approach to anticipatory systems at the urban scale (see Table 4.1). By 1992, sustainability became part of political agendas by adopting the Rio Declaration, leading governments to formulate sustainable urban mobility under an efficient transportation approach (Miralles-Guasch and Cebollada 2009; Montezuma 2003; Pozueta 2000). In 1994, the Aalborg Charter on sustainable European cities aimed to strengthen the implementation of strategies to reduce energy consumption, promote compact urban design, and foster economic development with social justice. Subsequently, the 2015 and 2030 Agenda and the SDGs consolidated three principal axes of urban mobility: social, economic, and environmental.

As an anticipation system, the economic dimension considers the efficiency of private transportation and maximizing public transportation benefits. Meanwhile, the social dimension includes issues of equity and access focused on the population of the

Table 4.1 The evolution of responses in urban mobility generation

Orientation	Reactive system	Anticipatory system
Approach	Transportation engineering	Sustainability
Vision	From Infrastructure	From mobility
Physical dimension	On supply and demand (flows and traffic)	Shape of travel (individual-space relationship)
Objectives	Connection with city spaces	Economic, social, and environmental sustainability
Hierarchy	On the vehicle	Inverted pyramid: On transport systems (Metro and buses), pedestrians and cyclists
Development oriented	Economic growth	Urban integration and transportation
Time and distance	Reduced	Accessibility and proximity

Source By the authors

lowest quintiles. On the other hand, the environmental dimension defines elements associated with reducing greenhouse gases and energy consumption and promoting clean public transportation systems as the central axis of mobility (Monzón 2005). These dimensions constitute an anticipation of urban planning that includes data, technology, and predictive models (Alcántara Vasconcellos 2020; Hernández 2017; Jirón 2017).

From these central axes of sustainable urban mobility, anticipation has been conceived in at least two primary dimensions: (1) urban development as a predictive model for land use planning and public transport, and (2) the prioritization of Metro systems in reconfiguring mobility.

4.2.1 Anticipation of Urban Development: Transit-Oriented Development

The first dimension of anticipation in sustainable urban mobility is associated with postulates on accessibility and equity, recently consolidated debates in the so-called Transit-Oriented Development (Rodriguez and Vergel-Tovar 2018). Accessibility involves considering satisfaction parameters, quality and coverage of transport systems (Metro and buses), road connectivity, physical and fare integration, and user satisfaction. This approach allows us to consider that accessibility, within an anticipatory system, considers elements related to how cities are developing and their growth patterns. In this sense, designing sustainable urban mobility requires thinking about future forms of the urban system and how to adapt cities to these scenarios to reduce or manage uncertainty. This necessity has compelled sustainable urban mobility to be conceived with a focus on the city's future locations of activities and infrastructure, the construction or expansion of transportation modes, and their connections with individuals (Hernández 2017; Montezuma 2003). The leading prospective research identifies the travel limitations of individuals based on predictive models of urban growth, considering the impacts on different population quintiles related to fares mixed with operating costs and the need for direct subsidies by governments to maintain accessibility (CAF 2017; Gutiérrez 2010).

Transit-Oriented Development (TOD) has been belatedly implemented in LAC cities. Curitiba in Brazil is considered one of the main cities in the region that has successfully integrated urban planning with mobility, particularly around BRT-type bus systems. For this city, planning must revolve around public transport as the central axis of mobility. Although it does not limit urban growth, it associates predictive capacity with responding to possible options for the evolution of urban forms through long-term policies that connect the sustainable road system with urban development. Quito, Bogota, Santiago de Chile, and Mexico City began constructing BRT systems in the 1990s and early 2000s to meet future mobility demands, while other cities sought to complement their Metro services.

However, these systems have been transformed along with the institutionalization of sustainable urban mobility policies. By the 1990s, the objective of BRT-type systems was associated with future demand, marking the first moment of anticipation. However, the exponential growth of LAC cities exceeded the capacity of these systems in some cities, forcing a transformation or update in response to the evolution of the urban system. By 2013, 63.6% of passengers were concentrated on BRT-type buses (Rodriguez and Vergel-Tovar 2013), intensifying intensified planning on the main road axes. For example, in Curitiba, the implementation of BRT buses was accompanied by the promotion of high-density land development and mixed uses, leading to the formulation of instruments related to urban management and public transport. This anticipation addressed uncertainty about urban development through long-term decision-making. Bogota and Quito have implicitly developed TOD models around BRT systems in the main transfer stations and terminals, promoting urban development, infrastructure, and accessibility (Vergel-Tovar 2019).

4.2.2 Metro Systems in the Reconfiguration of Mobility

Metro's systems have become a key element in urban mobility in terms of reaction and anticipation. In Latin America, the first Metro was built in 1913 in Buenos Aires, which by 2020 had six lines. While Brazilian cities have made remarkable progress implementing these systems, the latest Metro line extension was carried out in Brasilia, with 42 km of construction. Before the 1990s, Metro construction in Latin America was associated with the regional economic boom, coinciding with the economic growth of the main cities. However, this vision was not associated with a comprehensive understanding of mobility. In general, the provision of public transport services was considered poor quality, with low regulation by the State, resulting in quality, coverage, and road safety problems (Figueroa and Henry 1988). Through

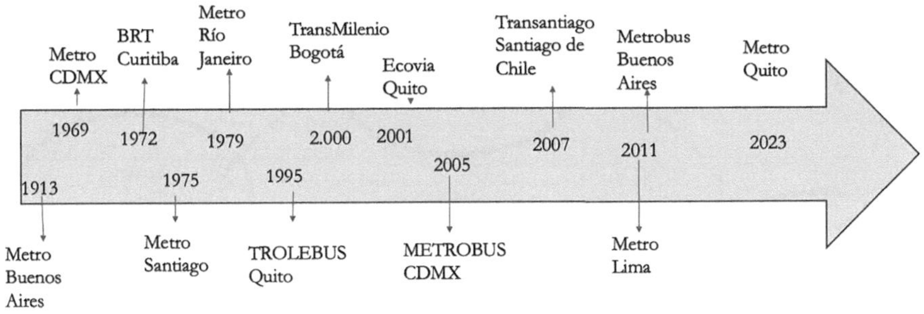

Fig. 4.1 Metro systems in LAC. *Source* By the authors

the concepts of comprehensiveness and sustainability, the Metro is considered an anticipatory system linked to environmental, socio-cultural, and economic axes (see Fig. 4.1).

The Metro de Santiago has been promoting MetroArte to link culture within the Metro stations through artistic works, dioramas, cultural diffusion, and value enhancements (De Grange 2010; Metro 2015, 2022). Artistic works have been installed inside some Metro stations, with the first one being installed in 1993 titled "Urban Interior," an oil on canvas by the artist Hernán Miranda. To date, 75 works have been exhibited, covering 11 thousand meters of permanent art. Additionally, a change in user behavior is being encouraged through the application, along with educational campaigns and the transparent Metro. That prioritizes open access to economic, technical, and financial information and regulatory frameworks.

The reconfiguration of mobility based on the subway as an anticipatory system is linked to behavioral changes, the positioning of infrastructure as a central point in travel, and the reduction of greenhouse gas emissions. The Santiago and Medellín subways, for example, have been promoting strategies such as MetroArte and Cultura Metro (Metro de Medellín 2023) through educational campaigns that consolidate the appropriation of the system while implementing changes in the use of energy, recycling, and waste from the train fleets. Santiago designed the "Huella Chile Program," which promotes the quantification of greenhouse gases within the subway infrastructure. Mexico City has promoted transparency strategies to ensure long-term policies strengthening the Metro system's predictive and transformative capabilities (Hidalgo 2022; Metro 2022; Ortuzar 1983).

In this sense, urban mobility as an anticipatory system is primarily framed in the institutionalization of long-term policies. That is associated with two main dimensions that promote the collective intelligence of cities (González-Arellano 2022) and encourage the application of predictive models in understanding present behaviors. Meanwhile, the predictive capacity dimension, environment, urban, socio-cultural, and economic development are associated with the system's capacity for transformation and its response to evolution in these areas. Considering this, the formulation of the TOD, the inclusion of climate change, social behavior change, and subsidies

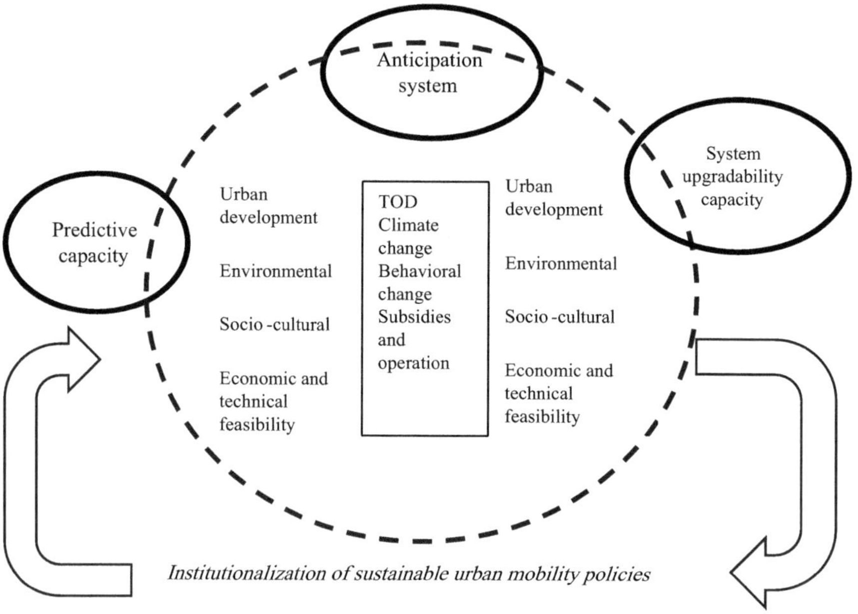

Fig. 4.2 Dimensions of urban mobility as an anticipatory system. *Source* Prepared by the authors

for operation constitute elements that enable anticipation in urban planning (See Fig. 4.2).

4.3 The Quito Metro as an Anticipatory System

With the declaration of the Law of Regime for the Metropolitan District of Quito (1993) and the decentralization reforms that Ecuador had been promoting since the 1990s to strengthen local governments, the municipality of Quito assumed a series of planning, regulation, and coordination competencies for various public services, including urban transportation. That led to a substantial improvement in public transport with the implementation of the Trolleybus in 1995 (the first BRT corridor), the modernization of the private bus fleet, and the construction of bus stops. This process was marked by a high level of conflict between the government, operating companies, and users associated with the problems of democratic governability of the emerging institutional framework of the metropolitan government of Quito (Chauvin 2006).

This background marks a change of vision from the reactive approach to transportation to the broad notion of sustainable mobility in the DMQ's policy agenda in the 2000s. Two factors marked this transformation: first, the urban development and environmental innovations incorporated by the Trolleybus project, which proposed a solution to mobility problems and was conceived as the axis of an integrated transportation system to structure a more equitable and environmentally sustainable city model. Second, the implementation of a series of instruments (laws and organizations), which underpinned a new institutional framework for urban mobility in terms of sustainable development, integrated into the Transportation Master Plan (2002) and specifically in the Mobility Master Plan 2009–2025 (Eguiguren 2015; Gordón 2011).

The formulation of urban mobility policy in the DMQ over the last three decades has been a process of innovation and political, social, and technological learning. Guided by the principles of sustainable development—environmental, social, and economic—various strategies have been implemented to assert mobility in the city as a fundamental right. Despite the limitations and shortcomings of the policies, progress is evident in areas such as improving the quality of public transportation and promoting non-motorized forms of transportation. These advancements highlight social actors' essential role, particularly activist groups and academia, whose participation and influence were crucial in reshaping the vision from a grassroots perspective, focusing on the needs of the most vulnerable. This process has redefined the culture and identity of mobility in Quito (Gordón 2011; Oleas and Albornoz 2015; Proaño 2012).

Historically, the modal distribution of trips in the city has been concentrated around 70% in public transportation (MDMQ 2015). Therefore, one of the central pillars of mobility policy has been developing an integrated mass transportation system. When the Trolleybus system faced expiration in the mid-2000s, an alternative proposal for a light rail system to enhance passenger capacity and travel speed was considered. However, this project did not materialize. Although the idea of constructing a Metro system had been previously discussed, it was dismissed at that time due to the high costs involved in the budget and the limitations of municipal management.

The proposal for a subway system resurfaced during the 2009 mayoral candidates' debate. Faced with the escalating demand to address the city's congestion issues, the construction of a mass transit system, specifically a Metro to replace the Trolleybus, gained significant traction in the political campaigns of the candidates. The mayor-elect embraced the Metro project as a key component of the new mobility policy,[1] and as the articulating element of a compact city model that optimizes land use and infrastructure in terms of rights. In the construction of the political agenda (Zahariadis 2007), the decision to promote a Metro system in Quito can be attributed to the convergence of three factors: a pronounced mobility problem that had garnered substantial attention in public opinion, and citizen demands; a redefinition of public

[1] The decision of the then mayor of Quito to discard BRT systems and promote the construction of a Metro can be explained by the logic of new public management, which seeks efficiency in the provision of the service. Additionally, it aligns with a discourse of equity and sustainability in urban mobility (Salar Khan 2016).

mobility policies based on sustainable development, advocating for the subway solution; and a political momentum reflecting consensus among various stakeholders to support the project.

Institutional restructuring was undertaken to implement Quito's new mobility policy, establishing the Secretariat of Mobility as the governing body for the policy and various operational bodies, including the company Metro de Quito. With technical support from the Metro Madrid Consortium, the conceptualization, studies, designs, construction, and equipping of the system for the First Line of the Quito Metro began in 2010. The project involved building 23 km of Metro tracks, 15 stations, and two multimodal stations in the city's north (El Labrador) and south (Quitumbe). The project was divided into two phases: Phase 1, which commenced in 2012, involved the construction of the two multimodal stations using municipal resources. Phase 2 encompassed the tunnel's construction, the remaining 13 stations, and all supporting infrastructure. The initial project budget was estimated at around US$1.5 billion, with funding expected from the municipality of Quito, the national government, and multilateral agencies (Oleas 2019).

The construction of the Quito Metro encountered various political, economic, and administrative challenges that led to delays in its completion compared to the initially established deadlines. The political instability within the DMQ's system, marked by frequent changes in mayors from different parties and movements, created a lack of continuity in city planning and policies. Each new mayor questioned the management of their predecessors, contributing to the obstruction and delay of the Metro project. Governance issues between the Metropolitan Council and mayors and tensions between the national government and the municipality further complicated matters. Financial complications arose when the national government withdrew the initially planned support, compelling the municipality to seek external resources for Phase 2 construction, thereby postponing the schedule. Challenges were encountered in financing, bidding, contracting, and operation processes. Despite these obstacles, 13 years after the project's initiation and under five municipal administrations, the Quito Metro was inaugurated in December 2023, costing US$2,017 million.

It has been noted that an anticipatory system is configured when the future is used to guide present decisions (Poli 2017). From this perspective, the Quito Metro is a relevant case to explore the anticipatory dimensions implicit in its design, both in terms of the future narrative on which the project has been delineated as the articulating axis of the sustainable mobility policy, and in the operational aspects of its implementation in different areas. That is because the Quito Metro project has just begun its operation. The design, construction, and execution processes developed in the last decade have a prospective and anticipatory character. We propose to analyze the anticipatory sense of the Quito Metro in three areas: urban development, environmental and climate change, and socio-cultural.

4.3.1 Urban Development as an Anticipatory System in Quito Metro

In its diagnosis of Quito's problems, the Metropolitan Development and Land Use Plan (PMDOT) highlights several critical issues regarding urban mobility. On the one hand, there is a constant growth in the number of vehicles, around 10% per year, which generates various adverse effects such as traffic congestion, carbon dioxide pollution, environmental deterioration, and health and quality of life problems for the inhabitants.[2] On the other hand, the supply of public transportation has failed to structure an integrated multimodal system. Ninety percent of this transportation is managed by 67 private companies, which operate disjointedly, and barely 10% of the transportation corresponds to the Integrated Public Transportation System (Trolleybus, Ecovía, North Central Corridor). Since 70% of the population travels by public transport, this generates coverage and accessibility problems (MDMQ 2021: 23, 31).

Considering this situation, the PMDOT (MDMQ 2021: 69) contemplates explicitly Strategic Objective 4: For a Connected Quito, aimed at providing reliable, quality, efficient, and safe mobility options. Under the principles of sustainability, vulnerability, and gender approaches, the policies of this objective propose planning and structuring a multimodal integrated mobility system following an urban development model based on multi-centralities. To this end, strategies and lines of action that discourage the use of private automobiles, increase non-motorized mobility, and consolidate sustainable metropolitan public transportation are proposed. In this sense, implementing the PLMQ and its extension to 2033 represents two central goals of this strategic objective.

Articulated from an integral perspective, a set of territorial systems configures the DMQ's Desired Territorial Model.[3] The city's growth has been conceived under a logic of polycentralities at different scales and influences, which takes advantage of the vocation of the territories and promotes a better distribution and territorial equity of infrastructure and services. The mobility and connectivity system structured around Quito is thus projected as the articulating axis of urban centralities and rural micro-regions, i.e., a multimodal system of Transport-Oriented Development, as a basis for promoting an inclusive, safe, resilient, and sustainable city (MDMQ 2021: 85,135). There is a sense of anticipation in the development model based on public transport, which serves the current public policy decision-making to reduce urban sprawl by consolidating centralities and more effective land use management.

Many mobility problems in the DMQ, such as vehicular congestion saturation, are directly related to imbalances in the spatial distribution of services, such as education (Astudillo-Ortega and Vásquez 2021). Hence, integrated transportation systems can

---

[2] In the case of the DMQ, market failures associated with mobility have led to, among other issues, health problems due to pollution and damage to heritage structures (Silva 2010).

[3] Environmental and Natural Risk System; Micro-regions System; Polycentric System; Mobility and Road Connectivity System; Heritage Areas System.

configure anticipatory development systems by appropriately targeting the location of metropolitan infrastructure and services, avoiding forced displacements.

The anticipatory character that the mobility and connectivity system has on the development of the DMQ is not only configured on a metropolitan dimension or scale, nor does it have a defined timeframe. The layout of the first line of the Quito Metro has generated expectations due to the impact of the different stops of the system in the sectors of the city where they are located. The location of the 15 stops was a matter of debate and dispute during its design, both in terms of the benefits and the negative externalities that the Metro circuit will have on the territory in the medium and long term.

Regarding urban implementation, two stops presented a high level of conflict between the residents of these sectors and the municipality. On the one hand, in the Solanda neighborhood in the south of the city, during the construction of the Metro stop, land subsidence problems affected the structure of several houses in the sector. This event opened a debate on the appropriateness of building a subway system in a territory with high physical vulnerability to natural hazards like the DMQ. On the other hand, some archaeological remains were found during the construction of the Plaza San Francisco stop in the city's historic center, which halted construction. At the same time, the heritage value of these vestiges was determined. Although the corresponding expertise determined that the ancient constructions found did not require special treatment, a broad academic discussion and protests from some citizen groups were generated regarding the danger of building the Metro stop in a historic sector declared a World Heritage Site. Not only was the intervention in an area of high physical fragility questioned but above all, the cultural transgression associated with its heritage status (Armijos 2021; Cabrera 2018; Plan 2016).

Based on the physical, economic, and social urban implications of the Metro's route in the territory, the system's design contemplated specific characteristics for each stop to counteract the impacts and transformations in the immediate surroundings. On the one hand, the infrastructure around the stations, through special urban-architectural programs and plans, is expected to promote urban renewal processes. However, there is also foresight regarding possible gentrification phenomena, i.e., displacement of the sector's original population and loss of identity, in high-density sectors that present physical deterioration, underutilization of land, and a deficit of infrastructure and public space (EPMMQ 2012).

The Quito Metro, a system that anticipates a series of urban transformations, mainly located in the immediate urban environment of the stops along the route, will directly impact the strategic decisions of several economic sectors. In the area of tourism, it is expected that the Metro will boost the flow of visitors to the city by increasing accessibility to sectors with tourist attractions such as the historic center (San Francisco station), but also access to sectors with a high supply of tourist accommodation and food services (El Ejido and La Alameda stations). Certainly, improving public transport in the city through a quality system can configure competitive advantages in strategic economic sectors such as tourism (Ávila 2017).

On the other hand, the anticipatory vision of the Quito Metro configures a series of urban development scenarios by modifying the current land use and occupation of those consolidated sectors through which the line passes. Depending on each typology's socio-spatial characteristics and territorial vocation, different growth hypotheses can be constructed regarding building and population density (see Table 4.2). The areas where type D stations are located have the most remarkable propensity for morphological transformation and demographic increase, owing to their commercial and public space dynamics. In a sector such as Jipijapa, located in the hyper center of the city, a high potential for urban development is projected in the medium and long term based on factors such as its level of accessibility, an average occupation of space for construction, and a tendency to change land use toward mixed activities, especially commerce. In the short term, the intensification of pedestrian flow at the Metro stop may induce a renewal of the urban infrastructure associated with pedestrian access to the area (Maldonado 2018).

The anticipatory vision of the Quito Metro has also been built in a utopian sense, that is, as an idealized projection of the functioning of the urban mobility system concerning the city's needs. While this vision is part of the narrative of an integrated multimodal system that provides accessibility and connectivity to most of the metropolitan district, this coverage has not materialized. The first line of the Quito Metro covers only part of the central urban area, and the multimodal system has problems consolidating due to conflicts in the public–private articulation of the transport management. That has meant that many district sectors, especially in the periphery, do not have access to quality public transport. An example is Calderón, a rural parish located in the north of the city, which has one of the highest demographic and urban growth rates in recent decades but was not considered within the Metro route. From a prescriptive and speculative sense, the need to extend the Metro line to new sectors has been discussed around this and other similar cases under an argument not only for mobility demand and urban development (Moreno 2020).

4.3.2 Environmental: Anticipatory Response to Climate Change

Many of the problems derived from the negative externalities of urban mobility in contemporary cities are related to environmental issues. The excessive circulation of motorized vehicles triggers the emission of polluting gases and high noise levels. The DMQ's vehicle fleet shows an annual growth rate of 5–10%, with a projection of 1,140,000 vehicles in 2030, and a motorization rate of around 192 vehicles per 1000 inhabitants, which could reach 450 in 2030. These values are among the highest in the region and explain the urban mobility problems experienced by the district due to motorized transport. On the one hand, there is an increasing difficulty in circulating and high levels of congestion, especially in the Hypercenter area where most activities are concentrated; travel times have increased by about 10% in the last decade, and the

Table 4.2 Types of Metro Quito stations by urban environment characteristics

Typology	Location	Characteristics	Development hypothesis
A Intermodal stations	Quitumbe Moran Valverde El Recreo La Magdalena El Labrador	Terminals connecting to the integral Mass Transportation System (SITM) with a high flow of user and commerce	Its urban environment is expected to increase its building density, mainly for purposes complementary to the transportation program
B Special stations	San Francisco Universidad Central Iñaquito	Due to their tourist, educational, commercial, and governmental characteristics, a particular study is recommended for each case	As they are in highly consolidated sectors, their functionality will directly relate to each sector's prevailing land use and occupation
C Administrative-Commercial stations	La Alameda El Ejido La Pradera La Carolina	Located within the hyper center of Quito, with intense commercial and administrative use, highly consolidated sectors where tall buildings predominate (six floors or more)	The intense flow of users to and from the stations is expected. Due to building consolidation, it is unlikely that new real estate developments will emerge to modify the environment
D Residential-Commercial stations	Solanda El Calzado Jipijapa	Residential sectors with a high percentage of sectoral commerce, where the predominant buildings are of medium and low height (less than six floors)	A possible re-densification of buildings, along with demographic densification and compaction around the stations

Source Based on Maldonado (2018), p. 20

average speed does not exceed 14 km/h. In environmental terms, because of heavy traffic and congestion, it has been determined that motorized transport, in general, is responsible for 56% of the carbon footprint in Quito (MDMQ 2015).

When faced with this problem, the DMQ's urban mobility policy has set, among its guiding principles, the preservation of the environment and the population's health. That has implied a change of vision toward sustainable mobility aligned with the Development Goals (SDG) as a response to the social, economic, and environmental impacts derived from the intensive use of motorized vehicles. Socially, the aim is for people and goods to move around efficiently and equitably. It is also expected to optimize travel times and operating costs to improve the urban economy. Environmentally, it is expected to promote transportation with renewable energy in the medium and long term. To this end, several strategies have been implemented, including vehicle controls to prevent polluting emissions[4]; promotion of non-motorized and pedestrian modes of transportation; incentives for alternative uses of private vehicles to reduce travel demand; and, mainly, the optimization and promotion of public transportation through an integrated multimodal system, where the Quito Metro[5] is the articulating axis of the system (MDMQ 2015; PMM 2009).

Environmental issues, in their broadest sense, have been on the political agenda since the 1970s, initially through specific plans (water, sewage, solid waste, and others) and later as a central axis of urban planning. As the environmental impacts associated with the development model have become more complex, new issues have been added to the agenda, including climate change. Faced with the imminent impact of climate phenomena in cities, since the early 2000s, the MDMQ assumed the responsibility and leadership to build a policy of mitigation and adaptation to climate change, which, after an important political and technical process, resulted in the Quito Climate Change Action Plan (Barrionuevo 2020). Although climate change is perceived as a problem managed from an environmental perspective, its causes and consequences are broader. Therefore, its articulation with other areas, such as sustainable mobility, is essential to understand the anticipatory nature of a system like the Metro.

The vision of the Action Plan is to make Quito a carbon–neutral and resilient territory in the face of climate change by 2050 through 19 actions related to governance, urban growth, land use, environmental services, and sustainable buildings, among others. Analyses of the DMQ situation have determined that the carbon footprint is caused by solid waste management energy consumption in the industrial, productive, and household sectors, agriculture in rural areas, land use change, and mainly urban mobility. In response to the mobility problem, the plan proposes three specific actions: (i) zero-emission public transportation, aimed at promoting the progressive replacement of conventional and public buses with clean technologies; (ii) integrated and efficient public transportation, to increase public transportation trips and reorganize routes and fares around the Metro as the central axis; (iii) active mobility, focused on increasing the network of urban cycle paths and supporting infrastructure for bicycles (MDMQ 2020).

Thus, the Quito Metro, conceived as a transversal axis of sustainable mobility, configures a system that anticipates the problems generated by climate change in at least two dimensions. Due to the Metro's accessibility characteristics, there is a real possibility of reducing the number of vehicles in transit, but also of optimizing

---

[4] In 2003, vehicle technical service centers were implemented for a mandatory annual inspection of private vehicles and twice a year for public and commercial transportation. However, the increase in pollution corresponds to the growth of the vehicle fleet, which mostly continues to use gasoline and diesel as fuels.

[5] The EPMMQ issued the Environmental Impact Study of the First Line of the Quito Metro (EPMMQ 2012). Beyond the technical details of the report, critical positions questioned the reductionist and fragmented vision of the city regarding environmental problems, suggesting specifying the scope of the impact in categories such as quality of life and the impact on biological communities (Aguilar et al. 2013).

travel times and encouraging the use of other non-motorized transportation alternatives such as bicycles, directly impacting citizens' health and quality of life. On the other hand, the energy used by this form of transportation is electrical, created by hydraulic sources. Therefore, the intensive use of this renewable energy, which does not require fossil fuels, will effectively mitigate the emission of polluting gases in the medium term. These would be local solutions to the global environmental impact produced by the negative externalities of the vehicle fleet. In this process, citizens have a fundamental role as agents of transformation toward sustainability. Thus, constructing environmental awareness, generating motivation, and empowering the population constitute social resources of an anticipatory character for Quito's Metro (MDMQ 2020).

4.3.3 Socio-Cultural Implications

The 2008 Constitution of Ecuador established a new vision of public action based on rights. That has implied that the design and implementation of public policies at different levels of government are structured in a binding manner on principles of equity, inclusion, and participation. Expressly, the notion of the right to the city has set a new course in territorial development, promoting urban planning based on sustainable and inclusive development (Pinto et al. 2023). Within this logic, DMQ's planning promotes, through its strategic objectives, a city of well-being and rights with equal opportunities for all citizens, especially for historically excluded and vulnerable groups. For this purpose, it is proposed to promote the exercise of rights by increasing access and quality of services and ensuring material conditions to reduce structural inequality gaps (MDMQ 2021).

Beyond its operational dimension, the DMQ's urban mobility policy is conceived as a critical sector for the district's progress. It promotes an inclusive, safe, and sustainable transportation and connectivity system that contributes to reducing inequity and promoting communities with a better quality of life. As part of the principles that guide the decision-making processes, there is a set of qualities that, from an anticipatory sense, focus on the social dimension of mobility in terms of rights. In this sense, accessibility guarantees adequate scope and coverage, as well as accessible costs and timely information. Equity and inclusion are focused on providing a service without discrimination of gender, age, capacity, or condition. Citizen participation seeks to guarantee a responsible inclusion of citizens within their competencies in planning mobility programs and projects (MDMQ 2015: 38).

The Quito Metro, as the axis of urban mobility, is an anticipatory element of the model of the right to the city model that guides the territorial planning of the metropolitan district. The transformative capacity of the Metro system, not only in terms of coverage and quality but also in terms of technological innovation, opens a window of opportunity in the medium and long term for the notion of sustainable mobility to consolidate and promote substantive changes in the city's structural inequities. Problems of socio-spatial segregation, urban dispersion, fragmentation, poor distribution of infrastructure and services, and excessive specialization of land use, among others, can be redefined in planning based on the advantages of accessibility and inclusion presented by the Metro.

From an anticipatory sense, the analysis of the impact of the first line of the Metro shows, on the one hand, that, in general, it will increase the accessibility of urban opportunities (Olarte 2021), depending on travel time and current levels of accessibility. On the other hand, a modest reduction of the accessibility gap is expected, with more visible benefits in long travel times (Quezada Larriva et al. 2023). However, it should be remembered that greater accessibility by connecting peripheries and centralities is not a measure that ensures the reduction of social exclusion in the city but needs to be integrated into a broader development model (Ortiz 2022).

As an anticipatory system, the Quito Metro has substantial implications for constructing a civic culture of citizen coexistence. Around the figure of the Metro, recognized as the most important urban infrastructure project in the city and the country, not only a normative dimension of the proper use of the service is being promoted, but, in a broader sense, a narrative of appropriation and respect toward common goods.[6] Certainly, the narratives configure anticipatory models, which simulate and stimulate key cognitive processes in the projection of the future (Lively 2017: 6). These strategies have been previously implemented in other municipal public transport systems such as the Trolleybus and Ecovia. For example, campaigns against the damage of stops and buses and the prevention of gender violence have been conducted, with relative results. That becomes important in a city like Quito, where urban mobility has historically been perceived as an aggressive, stressful, and constantly competitive space, inducing individual behavior of drivers and pedestrians to reproduce patterns of violence to adapt to the system and achieve the objective of mobility.[7] Although citizens know the rules governing mobility, compliance with them is partial, normalizing anarchy and impunity (Cáceres 2019).

In response to this situation, the MetroCultura strategy[8] has been conceived as a social management model based on principles of coexistence and participation,

---

[6] The cultural dimension of the Quito Metro had an important debate in September 2008, when a group of people broke into the Metro workshops' facilities and garages of the Metro to graffiti one of the cars. This act, categorized as vandalism and aggression to public property, generated a comprehensive discussion for and against the implications of criminalizing expressions of street art, but above all, about the new transportation system's cultural and public space dimensions (Sánchez 2021).

[7] The negative perception of urban mobility in Quito is expressed through low satisfaction (less than 50%) with the public transport service Linea Corridor Central Norte. That results in a high probability (57%) that users of this transport line will migrate to the new Metro system, indicating little identification with the current service (González 2021).

[8] The predecessor of this strategy is the Community Relations Plan included in the Environmental Management Plan (2012). Its objectives were: (i) to implement an inclusive relationship process with the different stakeholders; (ii) to promote a participatory process; (iii) to foster relationships

which is committed to the proper use and care of the transportation system. Above all, it aims to build a vision of Quito's future through the interaction of service users. To this end, the strategy proposes a system of values (respect, empathy, solidarity, tolerance, dialogue, responsibility) to govern the behavior of users in Metro facilities and spaces and to generate a sense of community among citizens. This vision has been summarized in the premise of "better users, better citizens" under three main approaches: Transportation and Public Health, Social Innovation, and Gender Equity and inclusion of priority groups. In operational terms, MetroCultura is implemented through two types of strategic plans: an information provision component on aspects related to the operation and use of the service and user service and an intervention component focused on training aspects, cultural consumption, and environmental awareness aimed at promoting new behaviors in system users (EPMMQ 2020). The strategy has a normative and prescriptive sense of regulation and control, and MetroCultura has significant communicational and pedagogical potential. It can be understood as an anticipatory system for the ethical transformation of individual and collective behavior in Quito's urban mobility.

4.4 Conclusions

Understanding sustainable mobility policies as an anticipatory system of development can be helpful in the decision-making process in the present. However, it is important not to project a reductionist narrative of a single preferred future but to consider the possibility of alternative images. That implies a systematic understanding of several possible futures (González-Arellano 2022). In this sense, subway systems are considered as anticipatory systems from different dimensions. Therefore, the case of the Quito Metro contributes to the debate of anticipatory systems around how the sustainable urban mobility policy was built and how it has been institutionalized implicitly in Quito's planning.

On the other hand, unlike the older subways created before thinking about sustainable mobility that has achieved greater consolidation of different strategies, the Quito Metro poses a substantive change in the mobility policy in the city, although still with challenges in institutional management and administration. Finally, the Metro, as an anticipatory system, provides different future scenarios in the present. Among these is a more integral reading of mobility, development, and culture-oriented to internalize the infrastructure within users and society.

These anticipatory policies of the Quito Metro stem from an articulation of anticipatory communities. In the construction and implementation process, a series of state and non-state actors with shared visions of a future and an anticipatory regime have converged. The case of Quito Metro reveals two types of communities. The first is related to a strategic anticipatory vision of political leadership, corresponding

---

based on trust, respect, and understanding; and (iv) to establish a dialogue relationship that prevents conflict situations during the construction and operation of the Metro (EPMMQ 2020).

to the vision of the local government, whose axis is sustainable development. This government promoted the project from a systemic perspective of the medium and long term, with interventions in the social and environmental spheres. It adopted a pragmatic and creative vision of the future.

The second community is configured from an anticipatory community of a managerial nature, centered around a new political situation designed to address the subway as a public transportation system. This community is anchored to a more operative vision of the local government, involving multilateral actors in the strategic management of the subway and in the construction process. This configuration establishes the subway as a transportation system operated by the Metro Company and other actors. However, a particular observation in the case of Quito is that once the project is concluded, it starts to take shape around the social actors. These actors contribute to building a Metro culture and new political leadership from the local government with a substantive development vision. Learning from the Quito Metro leads us to consider that communities are not exclusive to one policy area but function in response to the trajectory of decision-making.

References

Acosta A (2015) El Buen Vivir como alternativa al desarrollo. Algunas reflexiones económicas y no tan económicas. Política y Sociedad, 52(2). https://doi.org/10.5209/rev_poso.2015.v52.n2.45203

Aguilar C, Guerrero L, López E, Rodríguez M, Scarpacci M, Bustamante T, Paredes M, Perelman P (2013) Calidad de vida y comunidades biológicas: análisis del estudio de impacto ambiental del Metro en la Ciudad de Quito. Letras Verdes 13:41–52. www.flacsoandes.org/letrasverdes

Alcántara Vasconcellos E (2020) Equidad y sostenibilidad en la movilidad urbana de América Latina. In: Figueroa O, Valenzuela-Montes LM, Brasileiro A (eds) Desafíos del desarrollo urbano sostenible en el transporte y la movilidad (1st ed., vol 1). El Colegio Mexiquense A.C., pp 37–54

Armijos AM (2021) Imaginarios y discursos del patrimonio quiteño la Plaza de San Francisco y el Proyecto Metro de Quito [Tesis de maestría]. Universidad Andina Simón Bolivar

Astudillo-Ortega S, Vásquez A (2021) Detectando los desequilibrios dotacionales como generadores de problemas en la movilidad del Distrito Metropolitano de Quito. In: En Prceedings 3rd congress in geomatics engineering. Editorial Universitat Politécnica de Valéncia, pp 69–76. https://doi.org/10.4995/CiGeo2021.2021.12503

Avellaneda P (2017) Los proyectos de infraestructura y su relación con el ordenamiento territorial. In: Beuf A, Avellaneda P (eds) Ordenar los Territorios: Perspectivas y Criticas desde América Latina (1st ed., vol 1). Universidad de los Andes, pp 183–196

Ávila MS (2017) Análisis del impacto turístico del Metro de Quito en el centro histórico, una visión desde la geografía turística [Tesis de pregrado]. Pontificia Universidad Católica del Ecuador

Barrionuevo, M. de los Á (2020) La adopción de la agenda de política Pública ambiental urbana: el caso de la agenda de cambio climático en el Municipio del Distrito Metropolitano de Quito [Tesis de doctorado]. Facultad Latinoamericana de Ciencias Sociales

Bolay JC (2012) What sustainable development for the cities of the South? Urban issues for a third millennium. Int J Urban Sustain Dev 4(1):76–93. https://doi.org/10.1080/19463138.2011.626170

Cabrera S (2018) El patrimonio en disputa: la plaza vs el Metro de Quito (Paradiso Editores, vol. 1). Universidad Andina Simón Bolivar

Cáceres E (2019) Movilidad urbana en Quito: Análisis del funcionamiento del transporte público desde una perspectiva de la interacción cotidiana y la sociología de las emociones [Tesis de pregrado]. Pontificia Universidad Católica del Ecuador

CAF (2011) Desarrollo urbano y movilidad en América Latina. Banco de desarrollo de America Latina CAF

CAF (2017) Crecimiento urbano y acceso a oportunidades: un desafío para América Latina

Casado Izquierdo JM (2008) Estudios sobre Movilidad Cotidiana en México. Scripta Nova, XI I(273):741–798

Cebollada Á, Avellaneda P (2008) Equidad social en movilidad: reflexiones en torno a los casos de Barcelona y Lima. Scripta Nova, 12. https://raco.cat/index.php/ScriptaNova/article/view/115445

CEPAL (2018) Estudio Económico de América Latina y el Caribe 2018. Evolución de la inversión en América Latina y el Caribe: hechos estilizados, determinantes y desafíos de política (1st ed., vol 1). Naciones Unidas

CEPAL (2019) Estudio Económico de América Latina y el Caribe 2018. Evolución de la inversión en América Latina y el Caribe: hechos estilizados, determinantes y desafíos de política (1st ed., vol 1). Naciones Unidas

Chauvin JP (2006) Conflicto y gobierno local : el caso del transporte urbano en Quito [Tesis de Maestría, Facultad Latinoamericana de Ciencias Sociales]. http://repositorio.flacsoandes.edu.ec/handle/10469/424

Cosacov N, Di Virgilio MM (2018) Presentación del Dossier #10: Movilidades espaciales de la población y dinámicas metropolitanas en ciudades latinoamericanas. Quid 16(10):1–16

De Grange L (2010) El gran impacto del Metro. EURE 36(107):125–131

Eguiguren FJ (2015) ¿Caminando a la movilidad urbana sostenible?: desafíos de la implementación de políticas en la ciudad de Quito. [Tesis de maestría, Facultad Latinoamericana de Ciencias Sociales]. http://repositorio.flacsoandes.edu.ec/handle/10469/8420

EPMMQ (2012) Estudio de Impacto Ambiental de la primera línea del Metro de Quito

EPMMQ (2020) Metrocultura

Errazuriz M, Taddia A, Ríos R, Pérez J, Brennan P, Ortiz P (2017) Evolución de los Sistemas de Transporte Urbano en América Latina. Bid, p 67. https://doi.org/10.18235/0000828#sthash.yADhMxVk.dpuf

Estevan A, Sanz A (1996) Hacia la reconversión ecológica del transporte en España (1st ed., vol 1). Bakeaz

Figueroa O (2001) Políticas de desarrollo y políticas de transporte urbano. Coherencias y contradicciones. In: Carrión F (ed) La ciudad Construida. urbanismo en América latina (1st ed., vol 1). FLACSO, pp 377–391

Figueroa O, Henry E (1988) Diagnóstico de los Metros en América Latina. Revista Eure, XIV(42):7–17

González S (2021) Relación entre la percepción de la calidad del servicio del Corredor Central Norte de Quito y la probabilidad de migración al Metro. Neutrosphic Comput Mach Learn 16:82–95

González-Arellano S (2022) The future scenarios of cities: an analysis of their institutional construction. Int J Urban Reg Res 46(5):865–873. https://doi.org/10.1111/1468-2427.13103

Gordón S (2011) La movilidad sustentable en Quito: una visión desde los más vulnerables [Tesis de maestría, Facultad Latinoamericana de Ciencias Sociales]. www.flacsoandes.edu.ec

Gudynas E (2010) Desarrollo sostenible: una guía básica de conceptos y tendencias hacia otra economía. Otra Economía, IV(6):43–66. www.riless.org/otraeconomia

Gutiérrez A (2010) Movilidad, transporte y acceso: una renovación aplicada al ordenamiento territorial. Scripta Nova. Revista Electrónica de Geografía y Ciencias Sociales, 14. https://revistes.ub.edu/index.php/ScriptaNova/article/view/1736/1712

Gutiérrez A (2012)¿Qué es la movilidad? Elementos para (re) construir las definiciones básicas del campo del transporte. Bitacora 21(2):61–74. http://www.redalyc.org/articulo.oa?id=74826255011
Hammershøj LG (2017) Diagnosis of the times. Handbook of Anticipation. Springer International Publishing, pp 1–20. https://doi.org/10.1007/978-3-319-31737-3_64-1
Hernández D (2017) Transporte público, bienestar y desigualdad: cobertura y capacidad de pago en la ciudad de Montevideo. Revista De La CEPAL 122:165–184
Hernández D, Falavigna C (2016) Inmovilidad en la ciudad: el perfil del ciudadano "inmóvil" en Córdoba y Montevideo. Implicancias metodológicas y sustantivas. Universidad Católica Del Uruguay, pp 1–12.
Hidalgo R (2022) La trama del Metro. Patrones de transformación en torno a las estaciones del Metro de Santiago (Chile). Dearq 32:46–58. https://doi.org/10.18389/dearq32.2022.06
Jirón P (2017) Presentación. Movilidad Urbana y Género: experiencias latinoamericanas. Revista Transporte y Territorio, 16:1–8. https://doi.org/10.34096/rtt.i16.3600
Keivani R (2010) A review of the main challenges to urban sustainability. Int J Urban Sustain Dev 1(1–2):5–16. https://doi.org/10.1080/19463131003704213
Lazo A (2008) Transporte, movilidad y exclusión. El caso de Trasantiago en Chile. Scripta Nova, XII(270):741–798. https://www.ub.edu/geocrit/sn/sn-270/sn-270-45.htm
Levy A (1999) Urban morphology and the problem of the modern urban fabric: some questions for research. Urban Morphol 3(2):79–85
Lively G (2017) Anticipation and Narratology. Handbook of anticipation (1st ed., vol 1). Springer International Publishing, pp 1–20
Lupano JA, Sánchez RJ (2009) Políticas de movilidad urbana e infraestructura urbana de transporte. Documentos de Proyectos, 230(LC/W.230-P/E):65
Maldonado D (2018) Influencia del transporte y la movilidad en el desarrollo urbano de Quito, construcción del Metro de Quito, influencia urbana de la estación Jipijapa en el sector [Tesis de maestría]. Pontificia Universidad Católica del Ecuador
Manheim M (1979) Fundamentals of Transportation Systems Analysis (1st ed., vol 1). MIT Press, Cambridge, Mass
MDMQ (2015) *Municipio del distrito metropolitano de quito secretaría de movilidad visión estratégica de la movilidad*
MDMQ (2020). Plan de Acción de Cambio Climático de Quito
MDMQ (2021) Plan Metropolitano de Desarrollo y ordenamiento territorial
Metro de Medellín (2023) Relacionamiento con la comunidad. Cultura Metro. https://www.metrodemedellin.gov.co/cultura-metro/relacionamiento-con-la-comunidad/?hs_preview=WqETznOG-145687407978
Metro (2015) Metro de Santiago—Memoria Anual 2015. Memoria Anual. https://www.metro.cl/minisitio/memoria2015/ma-cap-1.php
Metro (2022) Nuestra História—Gobierno Corporativo—Metro de Santiago. Metro de Santiago. https://www.metro.cl/gobierno-corporativo/historia
Miller R, Poli R, Rossel P (2013) The discipline of anticipation: exploring key issues 1:4. http://www.costa22.dk/
Miralles-Guash C (2002) Ciudad y Transporte: El binomio Imperfecto (1st ed., vol 1). Editorial Ariel
Miralles-Guasch C, Cebollada À (2009) Movilidad Cotidiana y Sostenibilidad, una interpretación desde la geografía humana. Boletín de La A.G.E. 50:193–216
Montezuma R (2003) Ciudad y Transporte: La movilidad Urbana. In: Balbo M, Jordán R, Simioni D (eds.), La ciudad inclusiva (1st ed., vol 1). Naciones Unidas, CEPAL, p 313
Monzón A (2005) Gestión del transporte metropolitano. In: Rojas E, Cuadrado-Roura J, Fernández JM (eds.), Gobernar las metrópolis (1st ed., vol 1). Banco Interamericano de Desarrollo, pp 409–464
Moreno D (2020) La Política de movilidad en el Distrito metropolitano de Quito: El caso del Metro en Calderón [Tesis de pregrado]. UDLA

Olarte SH (2021) Which mobility variables are related to urban poverty? The case of Santiago De Chile And Quito. Archit City Environ 16(47). https://doi.org/10.5821/ace.16.47.9510
Oleas D (2019) ¿Cómo inciden los intereses y las instituciones en el cambio de agenda de la política de movilidad urbana? [Tesis de doctorado]. Facultad Latinoamericana de Ciencias Sociales
Oleas D, Albornoz MB (2015) La bicicleta y la transformación del espacio público en Quito (2003–2014). Letras Verdes 19:24–44. https://doi.org/10.17141/letrasverdes.19.2016.1817
Ortiz C (2022) Exclusión social e inmovilidad. Un análisis al diseño de la primera línea del Metro de Quito [Tesis de maestría]. Facultad Latinoamericana de Ciencias Sociales
Ortuzar S (1983) Santiago's Metro. Transport 113–116
Paz-Gómez D, Santelices M (2020) (In) capacidades de política en tiempos de Covid-19: comprendiendo las respuestas económicas de Colombia y Ecuador. Análisis Político, 100(1):72–91. https://doi.org/10.15446/anpol.v33n100.93362
Pinto V, Córdova M, Bell D (2023) The right to the city and contemporary housing policy in Ecuador. Hous Stud. https://doi.org/10.1080/02673037.2023.2180491
Plan V (2016) Los activistas que protestan contra las excavaciones del metro. https://Www.Planv.Com.Ec/Historias/Urbano/Activistas-Que-Protestan-Contra-Excavaciones-Del-Metro-Quito
PMM (2009) Plan Maestro de Movilidad para el Distrito Metropolitano de Quito 2009–2025
Poli R (2017) Introducing anticipation. In: Handbook of anticipation. Springer International Publishing, pp 1–14. https://doi.org/10.1007/978-3-319-31737-3_1-1
Pozueta J (2000) Movilidad y planeamiento sostenible : hacia una consideración inteligente del transporte y la movilidad en el planeamiento y en el diseño urbano (1st ed., vol 1). Instituto Juan de Herrera. https://books.google.com/books/about/Movilidad_y_planeamiento_sostenible.html?hl=es&id=GkmuAAAACAAJ
Proaño G (2012) Cultura "ciclera" en Quito, políticas de movilidad: estudio de caso ciclópolis y al sur en bici [Tesis de maestría]. Universidad Andina Simón Bolivar
Quezada Larriva A, Orellana D, Guerrero Balarezo ML, García JA, Cárdenas Villenas G, Osorio Guerrero P (2023) Impact of Quito's first Metro line on the accessibility to urban opportunities. J Transp Geogr 108. https://doi.org/10.1016/j.jtrangeo.2023.103548
Quintero J (2011) Inventarios Viales y Categorización de la Red Vial en Estudios de Ingeniería de Tránsito y Transporte. Revista De Ingeniería 20(30):65–77
Quintero-González J-R (2017) Del concepto de ingeniería de tránsito al de movilidad urbana sostenible. Ambiente y Desarrollo 21(40):57–72. https://doi.org/10.11144/JAVERIANA.AYD21-40.CITM
Rivas ME, Suárez-Alemán A, Serebrisky T (2019) Políticas de Transporte Urbano. In: BID. Banco Interamericano de Desarrollo, p 52
Rodriguez DA, Vergel-Tovar CE (2013) Sistemas de Transporte Publico Masivo Tipo BRT (Bus Rapid Transit) y Desarrollo Urbano en America Latina. Land Lines 16–24
Rodriguez D, Vergel-Tovar CE (2018) Urban development around bus rapid transit stops in seven cities in Latin America. J Urban 11(2):175–201. https://doi.org/10.1080/17549175.2017.1372507
Rodríguez Vignoli J (2008) Movilidad cotidiana, desigualdad social y segregación residencial en cuatro metrópolis de América Latina. Revista Eure, XXXIV(103):49–71
Salar Khan M (2016) Metro de Quito: better or bitter? Analysis of getting around Quito. PODIUM 29:93–105
Sánchez R (2021) Genealogía y criminalización de una subcultura urbana caso Metro de Quito, 2018 [Tesis de maestría]. Universidad Andina Simón Bolivar
Sheller M, Urry J (2006) The new mobilities paradigm. Environ Plan A 38(2):207–226. https://doi.org/10.1068/a37268
Silva J (2010) Movilidad urbana: análisis de los fallos de mercado ocasionados en el Distrito Metropolitano de Quito. Periodo 2009 [Tesis de pregrado]. Pontificia Universidad Católica del Ecuador
Vasconcellos EA (2015) Transporte urbano y movilidad (1st ed., vol 1). Universidad de San martín

Vergel-Tovar E (2019) ¿Es la densidad suficiente? Análisis de la relación entre la densidad poblacional y la demanda de pasajeros en los sistemas de transporte público masivo tipo BRT en Curitiba, Quito y Bogotá. Medio Ambiente y Urbanización 90(1):53–82 (30). IIED-América Latina

Winchester L (2006) Desafíos para el desarrollo sostenible de las ciudades en América Latina y El Caribe. Revista Eure XXXII:7–25

Zahariadis N (2007) The Multiple streams framework: structure, limitations, prospects. In: Sabatier PA (ed) Theories of the policy process. Westviwe Press, Boulder, pp 65–92

Chapter 5
"Se Va a Caer" (Patriarchy Will Fall): Care at a Possible Intersection Between Future Studies and Feminism

Daniela Losiggio and María Pozzio

"Y ahora que estamos juntas,
y ahora que sí nos ven,
Abajo el patriarcado
Se va a caer
Se va a caer;
Arriba el feminismo
Que va a vencer
Que va a vencer"
Popular and feminist song (Argentina 2018–2020)

Abstract In this chapter, the authors propose a dialogue between future studies and feminism. The feminist slogan "*Se va a caer*" (Patriarchy will fall) helps them think about this movement in terms of future narratives and the relationship between utopia-dystopia and uchronia in the context of threats and the rise of the ultra-right. Then, they propose some theoretical genealogies that allow them to conceptualize the notion of care and consider how practical proposals inspired by this concept can be regarded as anticipation.

Keywords Care · Future · Feminism · Patriarchy

D. Losiggio · M. Pozzio (✉)
CONICET, Universidad Nacional Arturo Jauretche, Buenos Aires, Argentina
e-mail: mpozzio@unaj.edu.ar

D. Losiggio
e-mail: dlosiggio@unaj.edu.ar

S. Gonzalez-Arellano and B. Gandlgruber (eds.), *Cities as Anticipatory Systems*, The Urban Book Series, https://doi.org/10.1007/978-3-031-78162-9_5

5.1 Introduction

In *Cuando el feminismo era mala palabra* (Tarducci et al. 2019), the authors describe the origins and actions of feminist demands during the years before and after the dictatorship in Argentina (1976–1983): at that time, these were isolated actions with low participation and, in general terms, promoted by urban and middle-class women. Since then and until today, on the 40th anniversary of democracy, feminisms have become plural and massive, as well as a significant actor in the political arena (we talk today about feminisms in the plural and, with it, we include popular feminisms, transfeminisms, indigenous women's movements, and lesbian movements, among others).

In the case of Argentina specifically, feminisms occupied the streets, starting with the first Ni Una Menos against femicide in 2015 and, later, with the *marea verde* (green tide), the name given to the massive vigils that accompanied the parliamentary debate on abortion in 2018 and 2020. The massification of feminisms constitutes a predominantly urban phenomenon, and this leads us to think about the city not as a given space—subsequently "occupied" by citizenship that fits its precepts—but as a system of networks that enable a "complex collective behavior" (see Chap. 2 Gonzalez-Arellano).

On the one hand, feminisms challenge the normative aspects of public infrastructure. Parks, public buildings, streets, and bus stops constitute power devices that regulate people circulation (Harvey 2007). Parallel, the glitter, the ornaments, the masks, the costumes, the colors, and the carnival—all of them feminist symbols in Argentina—call the sovereign exercise and manage to damage the ritualized and masculinized ways of inhabiting the city (Losiggio 2020; Losiggio and Taccetta 2022).

On the other hand, the massification of feminism expresses that cities have the capacity for self-organization, not necessarily arising from a control center, even when they can bend political wills and become institutionalized (Gonzalez-Arellano 2022). This phenomenon has attracted worldwide attention and required a fundamental component: social networks. Social networks extended the urban space throughout the country, generating a digital continuity in the city. Women and diversities in peri-urban or rural spaces participated in the demonstrations through the viralization of images and selfies showing outfits that responded to different slogans and hashtags.

Besides, this massification has also influenced public interest, agenda-setting, and, consequently, public policy. That is an unprecedented phenomenon. For the first time, public policies—including those aimed at public works and urban planning—incorporate the problems traditionally understood as private or intimate (sexual and reproductive health, gender identity, domestic violence).

In short, feminisms stresses the connotations of the city as the place of the public masculine (as opposed to the private feminine), showing that the city has the capacity to organize itself to exceed its precepts and devices. Finally, they manage to become institutionalized in different public policies. The song that became the score of the massive feminist demonstrations in Argentina in 2018, and says *Ahora que sí nos ven (…) abajo el patriarcado, se va a caer* (Now that they do see us, down with the patriarchy, it will fall), can be understood as the synthesis of the different anticipatory strategies of Argentine feminisms of the period 2015–2018; building a city in the present where that visibility is not only possible but desirable and safe; a more egalitarian city that materialize in actual campaigns against street harassment, murals that remember the victims of femicides, red benches to denounce violence against women, feminist taxi services to guarantee safe mobility, ad hoc daycare centers. In brief, as feminist urban planning organizations propose, each place of the patriarchal city must be transformed into a place of resistance by mapping, signaling, and sustaining organization networks. At the same time, the city must be the city of the young and the city of the old, your city and the global city.

In this article, we are especially interested in analyzing feminist anticipatory strategies around care. If we learned any lesson from the pandemic, it is that care activities are the basis for the sustainability of life. The idea of care represents an inversion of patriarchal and capitalist ways of understanding social existence: from the margins to the center, reproduction is an essential activity: housing, mobility, security, health, care of the land, food sovereignty, education over-exploitation, extractivism, pollution, and accumulation.

In times of the re-emergence of neoliberal economic policy proposals and the rise of new rights, statistics show us that it is still women (especially trans, indigenous, and rural women) and LGBTIQ + people who are affected the most by successive economic crises. Women and LGBTIQ + people continue to be those who mostly do domestic labor and those who embody the lowest-paid branches of employment in the market. Expressed in multiple proposals with significant impact on the urban scale and the world of work, the demand aims to develop public policies that make visible ("now that they do see us," says the song), de-privatize, and de-feminize care. Policies include creating care and parenting spaces, incorporating care issues—based on play—in school curricula, and paternity leave. The projection of public policies that can destroy the classic sexual division of labor—one of the main supports of patriarchy—can then be thought of as a form of anticipation toward a more egalitarian world.

Thus, throughout this article, we propose to analyze some notions of the future and anticipation that currently appear in feminist fiction and are much more effective than other genres in describing the threats that fall on women and diversities in the current context. Then, we will recover the genealogy of the sexual division of labor question to show the actual problem around care (some statistics and diagnoses). Finally, we will describe care public policies as forms of political anticipation and propose community care developed by women from popular neighborhoods in large cities as an example of this anticipation.

5.2 Feminism, Threats, and Dystopias

As Geneviève Liveley (2017) and María Mercedes Patrouilleau (2020) have stated, discourses and narratives provide a privileged setting for studying futures. Our motto, "The patriarchy will fall," can be understood through this lens. Now, if the score of the so-called green tide connotes optimism—the achievements of recent years in terms of rights support this emotion—it is also true that we write in the imminence of a global scenario with proposals that threaten to plebiscite rights, withdraw affirmative policies of parities and quotas, and defund sexual and reproductive health that can be victorious. So, what about the future?

The idea that the patriarchy will fall does not dictate the need for that event. The phrase constitutes an expression of desire founded on signs of the present, reliable information, scientific diagnoses, and collective networks alternative to the oppressive and extractivist ones. It aims to reopen the possibilities of the future and in some way expresses what anthropologist Arjun Appadurai (2015) calls the "politics of possibility": anticipation is not describing a destiny or reflecting a causality, but instead manifesting the possibility of doing something in the present to transform the future that appears obvious. That implies intervening in the temporal. Literature can do this by inventing a particular fictional hyperbole that cannot be accused of being tremendous because it belongs to the literary genre.

Utopias, uchronies, and dystopias bring us closer to critical concepts for the anticipatory exercise. Let us see one of our favorite dystopias, *The Handmaid's Tale*, Margaret Atwood's novel, transposed into a television series by Bruce Miller. Which concept of the future does this novel offer?

The term dystopia refers to fictions that outline a totalitarian society through a pernicious inversion of utopia, where our ideals are distorted and become nightmares. The dystopian narrative contains a critical and counter-hegemonic view of our contemporary society, implicitly including a warning. Margaret Atwood's novel describes the society of Gilead, shaken by mass infertility that causes a return to conservative religiosity. A caste of theocratic military and political leaders overthrows the democratic government and imposes a dictatorship that enslaves fertile women, prohibits women's public participation, and criminalizes gender diversity. For us, Gilead constitutes a warning: the phenomenon of generalized sterility is a metaphor for the control over our reproduction that we people with pregnant capacity have gained since the feminist struggle (the starting point of the so-called second wave). Indeed, radical theories about the origin of patriarchy revolve around this hypothesis: the will of men to control the reproductive power of women.

Furthermore, if Shulamith Firestone's utopia consisted of a dictatorship of the reproductive class (people with gestational capacity) and the retention of the means of reproduction, *The Handmaid's Tale* is presented as its opposite mirror. It plays with the idea of a reactionary reappropriation of that power by the patriarchy. Reading and circulating this text in a scenario like the current one allows us to diagnose the present critically and becomes a rational element of anticipation: indeed, we are threatened.

What are those threats? Several analysts have pointed out that the sexist and misogynistic discourse of the new rights in Latin America has the support of some sectors of the Pentecostal and Evangelical Churches (Bárcenas Barajas 2022), as well as Catholic organizations such as the Opus Dei (Vaggione 2005). In the debate on the right to abortion, these groups and their ideas gained visibility with a counter-symbol, the light blue scarves, and currently find expression in the political group of the Argentinian president, Javier Milei (Incaminato 2023). Within an ultra-liberal, free-market, anti-statist, denialist, and reactionary agenda, but also in its public interventions, his platform embodies a bellicose anti-feminist discourse with additives of the obscurantism typical of the radical right. The possibility of an electoral victory and a government by this political force openly threatens the lives of women and LGBTIQ + people. It refers to Atwood's dystopia as a latent horizon and reveals the necessary relationships between failed utopias, dystopias, and catastrophes (Engélibert and Guidée 2015). In the context of threat, dystopian fictions serve to make us understand the dangers of the future. The wish song "The patriarchy will fall" must not lose its meaning, but now it must be evoked with its disturbing reflection. If the essential thing about dystopia is to reflect a world that is distanced from the real one, but that serves to reveal it, these dystopian fictions reveal that patriarchy is not going to fall meekly without first giving furious kicks of reaction. In the context of our country, this reaction is embodied in the anti-feminist proposals of the new "libertarian" extreme right.

The Senegalese thinker Fellwine Sarr (2019) has been pointing out how the hypertrophy of the present is linked to two intentions of neoliberalism: to install the idea of no alternative (a future without promises) and the omnipresence of narratives where the only thinkable horizon is a posthuman future. Against this hypertrophy, he proposes "re-opening the future." The feminist slogan that predicts a future where "the patriarchy is going to fall" implies naming desire, leaving a time without promises, and investing the present and the future with meaning.

The idea of re-opening the future involves thinking about the possible challenges and future scenarios for women and the agendas of feminism (Hurley et al. 2008). In some way, the advances of the different feminisms of the world support this notion that "the future is feminist," but that, at the same time, depends on our actions. We are now interested in considering how we understand the care agenda proposed by feminisms in Argentina as a way of anticipation.

5.3 Care as Anticipation

5.3.1 *The Narrative of the Sexual Division of Labor*

However, what do we understand by care? To refer to the notion of care and understand its anticipatory possibilities, it is first necessary to analyze the Marxian narrative on the sexual division of labor and observe its consequences for the political struggles from the twentieth century to the present.

In the early writings of Karl Marx and Friedrich Engels and, eminently, in the great *Capital* (1867), it appears recurrently the idea of a natural division of labor, immemorial but still current, determined by sex and age, physical strength, and the different "roles in the sexual act" (Marx and Engels 2014, p. 21). With the development of the capitalist mode of production, they maintain, an increasingly specialized social division of labor emerged (famously described by Adam Smith before Marx). However, this social division does not replace the natural one. Women continue to be responsible for non-productive work (reproductive work). Then, in the factories, women and children work according to their physical conditions; in a word, less qualified jobs (Marx 2001).

Later, in *The Origin of the Family, Private Property and the State* (1884), Friedrich Engels tried to refine his understanding of the devaluation of female work and women in general. Its starting point is the thesis of Lewis H. Morgan and Johann J. Bachofen on archaic society and matriarchy, which were very much fashionable then and have been quite refuted today, especially in the field of anthropology. Based on these ideas, Engels maintains that there were extended families in "barbarian" kinship systems, parenting was shared, and monogamy did not exist. In a word, these were "communist homes" (2017: 60). In such contexts, descent could only be established through the maternal line (the only one genuinely verifiable), which contributed to granting women a robust social status.

With the invention of private property, men sought to guarantee better control over their offspring, appropriating them for the family economy. Maternal rights were then abolished, and women became "instruments of reproduction," those in charge of gestating, giving birth, caring for, and raising children; they were expelled from unproductive work (2017: 72). In this way, female oppression and the feminization of specific tasks were considered derivatives of the invention of private property. Consequently, Engels believed that with the communist revolution and the abolition of private property, female exploitation would disappear. Care, education, feeding, and cleaning would no longer constitute unpaid work.

Engels' theses were, to a certain extent, welcomed by feminists within communism, and for a long time, both demands overlapped. Although there have always been objections to the text, the most famous and fundamental one, dedicated to historical materialism, appears in the third chapter of *The Second Sex* by Simone de Beauvoir. Beauvoir maintains that Engels's narrative confirms the truths of historical materialism and leaves unexplained the singular character of the oppression it sought to criticize. The exploitation of women beyond the economic sphere (which is undeniable) must be read in its specificity. It is, above all, sexual: "In vain does rationalist materialism try to ignore the dramatic character of sexuality" (Beauvoir 2015, p. 121).

Feminist theory begins with an elementary premise: reproduction constitutes a historical fact inseparable from production. This proposal has been projected since the end of the 1960s. That year, the Argentine Isabel Larguía and the American John Dumoulin (2019), both committed to the Cuban Revolution, wrote in Havana a controversial and critical article that attacked the new Cuban regulations on the prohibition of the inclusion of women in certain kinds of employment. Larguía and Dumoulin (2019) invite us to think that the divorce between reproduction and production constitutes an ideological operation within Marxist theory and praxis. In the famous "The Unhappy Marriage between Feminism and Marxism" (Hartmann, 1981), Heidi Hartmann agrees with these ideas: if genders are distinguished, it is not because one group is capable of reproducing and the other is not. It is precisely the other way around: genders are the result of the need for the sexual division of labor, of one group alienating the other, both from material resources and from their sexuality.

Nancy Folbre and Ann Ferguson introduce another exciting idea in an unjustly forgotten text. The authors consider that female exploitation extends far beyond the domestic sphere. They refer to a "sexual-affective production," a notion through which it is possible to make the exploitation of women intelligible in a broad sense, transcend mere "procreation and raising children," and observe under the same lens "the provision of care, affection, and sexual satisfaction" at the social level (Ferguson and Folbre 1981: 314).

In neoliberalism and post-Fordism, social reproduction becomes a productive field for capital, which is converted into a source of profit. For this reason, we speak today of a "care crisis" (Kussy et al. 2023). Neoliberal economic policies damage what they should exploit but never destroy: the ability to feed, shelter, educate, and heal—the basis not only of capitalism but of existence itself—is put at risk for the entire population (Fraser et al. 2019).

These authors allow us to understand that for gender equality to exist economically, merely incorporating women into the labor market is insufficient. For true gender equality (but also class and racial equality) to exist, it is necessary to deconstruct patriarchal imaginaries, among which is the idea that women are the "natural" people in charge of care and "emotional" tasks and, on the other hand, these tasks must be understood as fundamental for the sustainability of life.

We live today in a systemic socio-ecological crisis directly linked to neoliberal capitalism and the damage that human beings themselves generate as a means of survival. This historical moment is called the Anthropocene, where humanity has become the cause of existence; no area of the environment has been affected by technology. In this context, feminism, together with environmentalism and communitarianism, has been proposing that it is necessary to project a new culture, called in Latin America "Buen Vivir" (Sumac Qamaña, in Aymara, and Sumak Kawsay, in Kichwa) that places in the center the caring, not only of other humans but also of the environment that allows the sustainability of life.

An inversion of the dominant narrative is thus proposed. What was previously understood as "natural reproduction" (outside sophisticated production, intellect, and culture) must be placed in the center as the most essential.

5.3.2 Feminist Economies and Care Ethics

Today, we call "care" the organization of material and symbolic resources that reproduce biological and social sustainability. They include self-care, direct care of other people (the interpersonal activity of care), the provision of the preconditions in which care is carried out (sanitation, food preparation), and the management of care (coordination of schedules, supervision of the housework). Feminized subjectivities have historically carried out these poorly paid or unpaid tasks and are also marked according to class and race (Tronto 1993, p. 112). The feminization of these tasks is neither natural nor casual: it has historically contributed to social non-recognition, invisibilization, and the symbolic—and therefore economic—devaluation of care jobs and those who carry them out. Care is thus either romanticized (in the figure of the mother) or vilified (the "dirty jobs" of cleaning or grooming elderly or sick people) but rarely recognized as valuable and value-producing work (Rodríguez Enríquez and Partenio 2020).

According to the CEPAL repository of information on time use in Latin America and the Caribbean (CEPAL, 2019a, b, 2021), women dedicate more than twice as much time to caregiving as men on our continent. The overload of unpaid work constitutes a barrier to participation in the remunerative labor market: women are the most informally employed, they suffer vertical segregation in public and private hierarchical organizations, and they receive much lower salaries than men (which is known as the gender pay gap). Furthermore, in the labor market, they occupy the branches of employment associated with tasks understood as "feminine" (such as domestic service, health, teaching, and social assistance). In feminist economics, these branches are called the "care sector." The countries with the highest concentration of women in this paid sector are Argentina (42.8%), Uruguay (38.4%), Chile (34.9%), and Brazil (33.7%).

This diagnosis allows us to observe more precisely women's greater economic dependence and relative impoverishment (OIT 2019). Authors such as Gago and Cavallero (2019) have shown that the recent financialization of the economy is mainly harmful to women, who are the ones who go into debt the most to support households.

During the COVID-19 pandemic, women became the primary caregivers and were responsible for the people who made up the new "risk population." With the pandemic, the care burden for women in the domestic sphere increased, and the inequality gap between women (which includes the cis, trans, urban, migrant, and native population) and men increased, too (MinCyt and MMGyD 2020). In the unprecedented exceptionality caused by the global pandemic, the word "care" acquired a relevance that cannot be understood as rhetoric. The pandemic highlighted that no decent life is possible without care, so it opened the possibility of problematizing it from an economic, ethical, and political perspective. It is necessary to deprive, de-feminize, and address the universality of care as an immediate policy goal. State institutions could consider equity strategies for their workers, incorporating care as a human right. Putting it this way means proposing a prospective intervention; it suggests lines of action in the present to outline a future that we understand is necessary.

5.4 Caring Cities

In Argentina, as we have already mentioned, women continue to be the prominent managers of household care and activities, as well as those in charge of "replacing collective infrastructure [for which the State takes no responsibility] in conditions of extreme precariousness" (soup kitchens, voluntary activities in schools, promoters of support mechanisms in situations of violence, health, and problematic consumption, etc.) (Gago and Cavallero 2019: 142).

In the popular neighborhoods of Argentina, women oversee the soup kitchens; if there is no work or the salary is not enough, they organize themselves. Under a roof, in their houses, back yards, or the streets, a pot, a fire, and whatever falls inside: noodles, vegetables, little bones, and meat, whatever they can find. The soup kitchens could serve as an indicator of the economic situation of the most impoverished sectors of Argentina. Every block has a pot, a soup kitchen, and a "glass of milk" because families have problems solving the most basic: food and the reproduction of life.

As we have been suggesting, the sexual division of labor is the basis of the feminization of care and its invisibility. Well, here, in the phenomenon of soup kitchens managed by women, we have an example of this and, at the same time, a sample of how care is becoming politicized: women take care; they resolve the reproduction of households, but not in the privacy of their homes, not in isolation, but with organization and visibility. Together and in the streets.

Liliana Albertazzi (2017) suggests that the micro-genesis of anticipation is the learning of living organisms from those experiences that have allowed them to survive. Well, in the peripheral neighborhoods of the big cities of Argentina, these learnings are historical: every time there was a crisis, women were in charge of "coming out to offer a pot." If, in the multiple actions developed in large cities, it is possible to discern anticipatory properties, do the experiences of the popular soup kitchens and what women do in the peripheral neighborhoods contain an idea of how to make a city that cares? How do you develop a care system's germ from a learned, palliative, reactive action?

Lately, the State has begun to develop institutional strategies for the deprivatization and defeminization of care. The thesis that the State is the origin of the sexual division of labor and, therefore, cannot be its remedy (Kussy et al. 2023) is not useful for understanding what is happening in countries like Argentina for several reasons. On the one hand, it responds to a societal perspective of the State, which observes it as a reflection of the market's interests. On the other hand, it assumes that it is a unitary and homogeneous actor, losing sight of the different logic of its agencies and the different rationalities that can manage it. In Argentina, in recent years, there have been exciting experiences developed by the Dirección Nacional de Economía, Igualdad y Género (National Direction of Economy, Equality and Gender), in the Ministerio de Desarrollo Social (Ministry of Social Development), and in some provincial ministries, such as that of Women and Gender Policies of Buenos Aires Province, to revert gender inequality in terms of wage gaps, and horizontal and vertical segregation. One of them, and probably the most important, is the inauguration of Early Childhood Spaces and Community Care Spaces around the national territory.

As part of the same context and political culture—where state capacities are built from social demands and state agencies are the subject of interpellation and demand to guarantee new rights, the urban social movements of Argentina have developed the project salary law for community cooks. Moreover, from these experiences, the State learned timid but relevant lessons.

In short, between what women do in the peripheral neighborhoods—the popular soup kitchens—the institutionalized proposals of the Ministries—national and provincial—and the salary project for community cooks, we can see several visions of the future that overlap and interact. As already stated in this book, "Futures emerge from a diversity of visions that are continually updated with the arrival and consolidation of events in the present, new information, or changes in values. However, amid this diversity, concrete futures are institutionalized, identifiable" (González Arellano, see Chap. 2 in this book). In these proposals, future scenarios coexist and complement each other as ways of imagining and promoting a shared space in cities. As they become institutionalized, they may converge with more hegemonic images of the future, such as those proposed by the UN Sustainable Development Goals. In the dialectic between the subaltern and the hegemonic, the capacities of these different actors develop. The big question is how it will be possible, in an adverse and threatening context, to project a more inclusive vision of a city.

5.5 Final Reflections

Throughout this article, we tried to put concepts of feminism into dialogue with future studies. We prefigure the present as a threat, and we have shown a counterpoint within the kind of desire that is expressed in the words "The patriarchy will fall." This slogan has taken over the streets of Argentina and other urban regions of the southern cone of Latin America. We have reviewed the theoretical contributions that allow us to discuss the sexual division of labor as a material structure that sustains patriarchy and to review the principles of other care ethics. We wanted this theoretical dialogue to allow us to understand the proposal for a comprehensive care policy as a system of anticipation. If these are action devices, intervention proposals that encourage future scenarios, undoubtedly the policies and institutions that implement measures of this type not only encourage an idea of the future (that idea that reopened feminism in a moment of hypertrophy of the present and the end of emancipatory narratives) but instead push it, in material terms, also in a socio-affective and symbolic way.

The proposed dialogue also sheds light on the different scenarios and threats that attack the women's and LTBGI + movements: on acquired rights, that is, on the present, but also, and above all, on the future. In this sense, the political proposal that places care at the center must not only be understood as an anticipation of these threats but also as a way of appropriating and giving concrete and utopian meaning to the desired, imagined times, to the intimate way in which past and present are linked, and above all, to the time to come. Every time a community cook in a soup kitchen prepares food for a child, she reproduces his existence. She has faith in the future of that child, her community, and the generations to come while, at the same time, updating a collective form of care that is an ethical message and a supply for State policies. That is, it establishes a priority and a form of action to be recognized and financed by the Comprehensive Care System for which it is fighting and which must consider the salary of those who care, in this case, cooking.

Every time a community cook prepares and delivers food, she makes that soup kitchen a space of care, a piece of home in the neighborhood. With that daily act, she citizens that corner, by definition private and feminine, which is the kitchen (and from where comes the Greek notion of *oikos*, hearth, the house, the fire that must be kept burning). That means the anticipation is carried out and takes shape in a social and, in this case, urban space.

If the city is a space for promoting new futures, these visions of shared spaces, of cities that extend the notion of care to the public sphere, can serve as inspiration. The literacy future works propose how the unimagined—via technological development, imagination, and science fiction—enters the agenda of urban policies (Gonzalez-Arellano 2014). Here, we tried to expand that possibility to understand how collective intelligence is developed not in the imagined nor the unimagined but in the unseen, the historically invisible.

This book has invited us to think of cities as anticipatory systems; then, the dialogue we have carried out should allow us to imagine and anticipate future cities, cities that put care at the center. A city that cares is a home. A city that, like the invisible cities written by Italo Calvino, is all cities. Furthermore, stop being invisible.

References

Albertazzi L (2017) Microgenesis of anticipation: windowing the present. In: Poli (eds) Handbook of anticipation. Springer International Publishing. https://doi.org/10.1007/978-3-319-31737-3_7-1

Appadurai A (2015) El futuro como hecho cultural: ensayo sobre la condición global. Fondo de Cultura Económica, Buenos Aires

Bárcenas Barajas K (coord) (2022) Movimientos antigénero en América Latina. Cartografías del neoconservadurismo. Ciudad de México, UNAM-IIS

Beauvoir de S (2015) El segundo sexo. Cátedra, Valencia

Engélibert JP, Guidée R (2015) Actualité de l'utopie. Utopie et catastrophe: revers et renaissances de l'utopie (xvie-xxie siècle) [en línea]. Rennes: Presses universitaires de Rennes, 2015 (generado el 15 septembre 2023). Disponible en Internet: <http://books.openedition.org/pur/177786>. ISBN: 9782753591592. https://doi.org/10.4000/books.pur.177786
Engels F (2017) El origen de la familia, la propiedad privada y el Estado. Akal, Madrid
Ferguson A, Folbre N (1981) The unhappy marriage of patriarchy and capitalism. In: Sargent L (ed) Women and revolution. Black Rose Books, Montreal, pp 313–338
Fraser N, Bhattacharya T, Arruzza C (2019) Manifiesto de un feminismo para el 99%. Herder, Madrid
Gago V, Cavallero L (2019) Una lectura feminista de la deuda. Vivas libres y desendeudadas nos queremos. Fundación Rosa Luxemburgo, Buenos Aires
González-Arellano S (2022) The future scenarios of cities: an analysis of their institutional construction. Int J Urban Reg Res 46(5):865–873. https://doi.org/10.1111/1468-2427.13103
Hartmann H (1981) The unhappy marriage of marxism and feminism. Towards a more progressive union. In: Sargent L (ed) Women and revolution. Black Rose Books, Montreal, pp 1–42
Harvey D (2007) Urbanismo y desigualdad social. Siglo XXI, Buenos Aires
Hurley K et al (2008) Future studies and feminism. Conversation. Futures 40. Elsevier, pp 388.407. https://doi.org/10.1016/j.futures.2007.08.004
Incaminato N (2023) El género en disputa: el voto libertario y el post-feminismo de las Pibas de Milei. 25 de junio de 2023, Sección Opinión en Diario.ar, disponible en: https://www.eldiarioar.com/opinion/genero-disputa-voto-libertario-post-feminismo-pibas-milei_129_10323102.html
Kussy A, Palomera D, Silver D (2023) The caring city? A critical reflection on barcelona's municipal experiments in care and the commons. Urban Studies 60:2036–2053
Larguía I, Dumoulin J (2019) Hacia una concepción científica de la emancipación de la mujer. Desde la Cuba revolucionaria: feminismo y marxismo en la obra de Isabel Larguía y John Dumoulin. comps. Mabel Bellucci y Emmanuel Theumer. Buenos Aires: CLACSO, pp 117–120
Liveley G (2017) Anticipation and narratology. In: Poli (ed) Handbook of anticipation. Springer International Publishing. https://doi.org/10.1007/978-3-319-31737-3_7-1
Losiggio D (2020) Universal y afectiva. La esfera pública en el pensamiento político feminista. Las Torres De Lucca 17:139–165
Losiggio D, Taccetta N (2022) The virtual museum and female joy in Argentina: an exploration of the revolt. Intérmedialités. Históire et théorie des arts, des lettres et des techniques, p 38
Marx K (2001) El Capital. Fondo de Cultura Económica, México
Marx K, Engels F (2014) La ideología alemana. Akal, Madrid
Patrouilleau MM (2020) Epistemología y crítica de la prospectiva y los estudios del futuro. Una visión desde América Latina. In: Patrouilleau M, Albarracín Deker J (coords.) (2022). Prospectiva y estudios del futuro: Epistemologías y experiencias en América Latina. CIDES-UMSA, La Paz. En Memoria Académica. Disponible en: https:www.memoria.fahce.unlp.edu.ar/libros/pm.5423/pm.5423.pdf
Rodríguez Enríquez C, Partenio F (2020) Sostenibilidad de la vida desde la perspectiva de la Economía Feminista. Madreselva, Buenos Aires
Sarr F (2019) Reouvrir les futurs. In: Achille M, Felwine S (dir.) Les ateliers de la pensée. Politique des temps: imaginer les devenirs africains. Dakar, Phillipe Rey eds
Tarducci M, Trebisacce C, Grammático K (2019) Cuando el feminismo era mala palabra. Algunas experiencias del feminismo porteño. Espacio editorial, Buenos Aires
Tronto J (1993) Moral boundaries. Routledge, Nueva York y Londres
Vaggione JM (2005) Entre reactivos y disidentes. Desandando las fronteras entre lo religioso y lo secular. Defensa de los derechos sexuales en contextos fundamentalistas, ed. Católicas por el Derecho a Decidir. CDD-Córdoba, IGLHRC, Buenos Aires

Informs

CEPAL (2019a) La autonomía de las mujeres en escenarios económicos cambiantes
CEPAL (2019b) Oportunidades y desafíos para la autonomía de las mujeres en el futuro escenario del trabajo
CEPAL (2021) Repositorio de información sobre uso del tiempo de América Latina y el Caribe
MinCyt/CONICET, MMGyD (2020) Diagnóstico de la situación de las mujeres rurales y urbanas, y disidencias en el contexto de COVID-19. Disponible en: https://www.conicet.gov.ar/wp-content/uploads/resumen_ejecutivo_mujeres_y_covid_-_mincyt-conicet_-_mingen.pdf
OIT (2019) Panorama laboral en América Latina

Chapter 6
Inclusive Urban Health Systems and Universal Well-Being: Mirage or a Real Future?

Bikramaditya K. Choudhary

Abstract The survival of the human race refers to the living humans dependent on their health and health systems, including preventive and curative measures. The contemporary shift toward urban settlement as the place of residence makes it imperative to understand the intricacies of urban spaces and their liveability. Cities as a healthy system and economy as a system of cities are often considered synonymous, though they are quite a contra in neoliberal regimes. The people are left to take care of their health in large cities, including occupational health and trauma cases, partly due to the economic processes that make the city. The idea of an "anticipatory city" is rightly posed as it holds the future. It is essential to point out that many individuals and institutions have answers for the future without knowing the problems; they have a vision for an unknown destination. This paper will conceptualize important questions for the larger masses destined to live in oblivion as their voices have been silenced in the loud noise of the economic successes of existing and emerging "great powers." Based on the discursive contextualization, I conclude that the core challenge for the future city is the tacit fortification of inequality on different counts and the unabashed defense of increasing consumption.

Keywords Diseases · Preventive health · Neoliberal economy

6.1 Introduction

It has been established more than once that there is no relationship between collective human achievement and the common good. On the one hand, cities are the real future of humanity in terms of preferred forms of settlement; at the same time, cities are termed "nightmares" or "dystopias" (Ameel 2016; Latham and Hicks 2014; Sulzbach 2013). The desirable utopia or contra to it can be conceptualized in multiple ways, including the cultural amalgamation of multiple stands of social practices, the cultural

B. K. Choudhary (✉)
CSRD/SSS, Jawaharlal Nehru University, New Delhi, India
e-mail: bkcjnu@gmail.com

S. Gonzalez-Arellano and B. Gandlgruber (eds.), *Cities as Anticipatory Systems*, The Urban Book Series, https://doi.org/10.1007/978-3-031-78162-9_6

conflicts on the same basis, the higher levels of living standards and contra to the same, and so on. Cities are a certain kind of spatial fix, a set of spatiality produced filled with ideologies, actions, conflicts, and negotiations.

Soja (2000: 6) argues that the processes of producing spatiality or "making geographies" begin with the body, the construction, and the performance of the self, the human subject, as a distinctively spatial entity involved in a complex relationship with our surroundings. On the one hand, our actions and thoughts shape the spaces around us, but at the same time, the larger collectively or socially produced spaces and places within which we live also shape our actions and thoughts. Production and reproduction of spaces shape places as they get identified, creating class- and class-based place-making. Class-based place-making, in return, creates an exclusionary cityscape with restrictions and the formation of "gated communities" and "ghettos" simultaneously within the same city. Cities in the contemporary world are utopia and dystopia at the same time, based on who is in charge of a particular space. The questions of power and marginalization also become factors in considering cities or parts of cities as either category. The advent, persistence, and continuance of the current pandemic established once again that cities can survive or perish together.

Countries like India, except during the second peak, considered COVID-19 a "disease of the city" as major mortality was reported from the large cities and remained concentrated in the large cities as it has a "long tail." It is important to note that India still (until the last census) has most people living in rural areas. During the pandemic's peak, millions of urban workers returned to villages to save themselves from the wrath of the disease and consequent "lockdown." The sufferers of the pandemic are mostly poor, as most poor households became dependent on government support to fulfill their basic needs, whereas, during the same period, large business corporations reported unexpectedly faster growth in their wealth and income. For those who view cities through a dystopian lens, the ongoing COVID-19 pandemic, with people fleeing urban areas, is another stark illustration. It raises the question: Can we envision and plan our cities more optimally to address such challenges? Can we anticipate the future? Can such pandemics be better managed? Can the sufferers of this pandemic (mainly poor and migrant workers) not be the sufferers of the next pandemic? The questions bring another set of questions, including vulnerability to disaster, inequality in terms of being affected, the nature and extent of resilience, and the possibility of a "fractured future."

We know that the distribution of health risks is uneven, as is the access to healthcare facilities. The existence and persistence of such inequalities can be seen globally when comparing life expectancy at birth in Japan (83.7 years) and Sierra Leon (50.1 years). Similar inequalities across gender and economic quintiles are visible nationally in rich and poor countries. There is a 20-year gap in male life expectancy between Glasgow's richest and poorest areas (Haining 2021). India is not an exception, and in any study, including in everyday newspaper reporting, one can come across glaring inequalities between men and women or rich and poor (Asaria et al. 2019). On top of that, intersectionalities across caste, gender, and economic categories make things complex. What is expected of the health policy of a democratic welfare government is a clear articulation to contain existing inequalities to move toward a healthier, more inclusive society.

More often than not, the issue related to health care raises the question of equity, especially in poorer countries. The history of human civilization is also the history of fighting diseases successfully and unsuccessfully. From Plague to Ebola, the relative success of human efforts has been varied across geographies at certain times. A partial explanation for such varied success in dealing with the "health issues" draws our attention to states' varying nature and priorities. The role of state and public policy has entered the debate since the existence of the "nation-state," especially the articulation of the "welfare state." The spatial dimension is critical in developing societies where healthcare institutions are concentrated in large urban areas and major diseases are primarily concentrated in rural areas or urban slums. Socio-cultural hierarchies and the varying purchasing power of individual households add to the complexity of health care and disease prevalence in these societies. The prevalence of disease also has roots in differentiated vaccination and varying vulnerabilities across the layers and strata of the population. It is, therefore, pertinent to discuss the nature of economic opportunities and differentiated vulnerabilities rooted in the socioeconomic conditions of individuals and families. The following sections of the paper will discuss the urban healthcare system, followed by healthcare facilities in Delhi. The last section will discuss the outlined conditions impacting cities and their future.

6.2 Inequality, Informalities, and Vulnerabilities in Delhi-NCR

Urban spaces are known for visible and invisible concentration and deprivation. Some thrive in cities at the cost of millions in countries like India. The metropolitan cities of India are known for their vibrant business environment, sprawling shopping plazas, international standard superspecialist health facilities, grand venues for business meetings, and so on. At the same time, these cities are also known for their vast homeless population, large slums, poor sanitation, unliveable environmental conditions, excessive pollution, scarcity of water, unsafe city transport, and so on. The rising income and galloping income gap move almost in a linear manner. Informality is the key constituent of these cities, whether in planning, the job market, land tenure, or housing. With that being acknowledged, urban regions have been a preferred destination for improving the standard of living for millions of underprivileged people in emerging countries. This is especially important because the villages that used to be their traditional homelands face economic stagnation.

The deteriorating condition of people living in rural areas leads to a heavy influx of migrants to urban areas. The growing inequality among the people and the growing inequality in the profit-making capabilities of urban processes have promoted urban informality through sub-contracting. A significant trend in growing inequality can be seen through the profit-making processes of different sectors. The high-profit industries dominate major urban economies. They have increased the price of space and land in the big cities (Sassen 2015). Low-income people can hardly afford the space, and hence, we witness high-density slums like Dharavi in Mumbai and scores of homeless people across Indian cities.

The informal settlement is described as a built environment that is either a slum, illegal or legal tenure (Wekesa and Styen 2011), or squatters of different kinds. Three interrelated characteristics, viz., physical, social, and legal, define the informal settlement (Wekesa and Styen 2011). Due to being an illegal settlement, the physical characteristics of such a settlement include the lack of adequate infrastructure like roads, electricity, water supply, toilets, schools, health centers, and markets. The people living in informal settlements mainly belong to a low-income group or are daily wage earners working in different unorganized sectors of the economy. The residents of these settlements lack ownership of the land on which they built their abode. The land people squat on is mostly government, public, or marginal land parcels like railway tracks, marshy lands, and more.

These settlements lack basic amenities and have improper housing, leading to vector-borne disease and making the residents vulnerable to communicable diseases. In developing countries, the proliferation of informal settlements results from many actors and processes, including speculative urban land markets. On the one hand, the speculative land market allows real estate actors to own the city while making houses out of reach for millions of families (Bhagia and Bose 2023; Sassen 2015). Private entities own more buildings, and the conventional building of houses has become beyond the reach of the urban poor (Wekesa and Styen 2011). The proliferation of informal settlements is witnessed due to market failure and public policy failure under globalization. It is well documented that, citing a lack of funds, the local government and municipality have abandoned providing proper infrastructure and basic amenities to these settlements. The land cost is so high that it is difficult for the average individual to build houses, which leads to a proliferation of slums and squatters.

In the twenty-first century, urban growth is increasing in the developing world, but city-related theories remain confined to the developed world. The future of urban was thought to lie in Third World cities like Mumbai and Rio de Janeiro (Messey 2001; Roy 2005); urban transformations in the developing world remained particularly associated with squatters. It is seen as a generalized mode of metropolitan urbanization. The idea of distributive justice in urban planning remained in writing and scholarship (Roy 2005) at best, and inequality and injustices continue to perpetuate on people unhindered (Srivastava 2015; Choudhary 2019; Saroj et al. 2020).

Informal hypergrowth of the city is considered (Roy 2005) as one of the causes of growing informality; however, the classical study of Engels established that the same forces that generate squatter in one place will produce it elsewhere unless the process of urbanization that is based on accumulation is not changed (Choudhary 2005). In India, most of the urbanization results from urban informality, with no provision of adequate infrastructure and a weak municipal governance structure. Most of these settlements are developed along the major transportation routes within the metropolitan boundaries. Informality is a different type of market where affordability accrues through the absence of formal planning and regulation. These are complex phenomena of legality and illegality where squatter settlements grow through land invasion (Roy 2005). Informality indicates to whom this belongs and who has the right to a city (Harvey 2003). The right to include and exclude plays in a manner that is influenced and questioned diabolically (Bloom 2004). Delhi accounts for 14.6% of the population living in informal settlements (Census of India 2011). A severe issue in Delhi is the lack of affordable housing for the lower and middle classes. Slums and Unplanned Housing: In 2014, the Delhi government assessed slums and JJ colonies, estimating that 0.33 million households, or around 1.7 million people, lived in these settlements, making up roughly 10% of Delhi's total population. The lack of housing, together with the miserable living conditions of the homeless, the abundance of slums, including unauthorized colonies and Jhuggi-Jhopri (squatter) settlements, the constant influx of people into the city, etc., force people to live in substandard conditions with little or no access to essential services and inadequate housing (Economic Survey Report 2022–23). Informality in housing and land ownership means frequent displacement within the city. Delhi has witnessed impoverishment due to displacement in many phases. The city witnessed demolition before the Asian Games in the early 1980s; in the 1990s, rapid urban construction witnessed large-scale displacement; and in 2010, the largest displacement was witnessed in Delhi during the Common Wealth Games. There were slum policies adopted or introduced with some conditions for their eviction in the 1990s, but the judicial intervention undermined the policy to a large extent (Bhan 2009). The evicted often lose their claims, as revealed in studies such as the Almitra Patel Case (2005) and the Okhla Factory Owners Association Case (2002), highlighting the state's failure to fulfill its obligation to provide alternative resettlement for evicted families in these situations (Dupont 2000; Ghertner 2014).

6.3 Access to Basic Services and HealthScape

Urban water supply systems are vital in ensuring access to safe drinking water for millions in different cities. The cities of the global south are known for having multiple waterscapes. In this broad primary circuit, urban local bodies supply water, and people living in the rest of the city spaces are left to fend for themselves. At times, water vendors, both formal and informal, make water available to them. The issues of water supply, access to clean water, and distribution systems in the cities have become central to discussions on improving access to safe water for millions in the Global South, like India (Sarkar and Choudhary 2020). The inadequacy of water supply in Indian cities, especially across informal settlements with precarious tenure status, such as slums and unauthorized colonies, is often highlighted (Dutta and Chaudhuri 2020; Ranganathan 2014; Björkman 2015). Conditions of water inadequacy across planned neighborhoods for the poor (resettlement and rehabilitation colonies) have a similar fate, as either they depend on vendors or, sometimes, on water ATMs in cities like Delhi (Sarkar and Choudhary 2020). The gap between water demand and supply continues to exist despite some augmentation efforts by the local bodies to supply water.

Delhi's National Capital Territory (NCT) has always had difficulties dealing with water, as the city has hardly access to groundwater, often 40 m below the surface. Further, the Delhi metropolitan limits contain the most polluted stretch of the Yamuna River. Even during monsoons, urban flooding in the metropolis, even with low rainfall, is commonly witnessed as faulty drainage and unplanned construction in the *laldora* (village boundary) area. Ensuring that citizens have access to safe, adequate, and affordable portable drinkable water and hygienic sanitation has to be a fundamental public service the government must ensure for the well-being of citizens. As per the Master Plan of Delhi 2021 formulated by DDA, the proposed water requirement is 80 gallons per capita per day (GPCD), of which 50 GPCD are for domestic uses. Domestic uses comprise 30 GPCD portable drinking water and 20 GPCD nonportable. The demand for water in Delhi based on the norm is 60 GPCD by 2021 is 1260 MGD to meet the requirement of the population of about 21 million people, whereas the installed capacity of water plants has increased from 810 MGD in 2009 to 943 MGD in 2022. There is still a more than 300 MGD gap, and the poor face the brunt.

New Delhi's sprawling suburb Gurugram, developed as a city to be built and serviced by private real estate developers, exemplifies the necessity of considering public transport and road safety as a public health concern. With a sparse public transport network, the city is home to approximately 2 million people and is served by a paltry network of 20 routes and 142 buses. This leaves large swathes of the city, at best, underserviced and, at worst, unserved. The Rapid Metro Gurgaon, another privately conceived and built project, serves the commercial and new sectors of the city and is notorious for its poor network spread of 12.85 km and only 11 stations. In the absence of any public transport network worth its name, private cars are the preferred mode of transportation, while the urban poor are mainly at the mercy of shared private autos and cabs ferrying people throughout the city and beyond. The Statistical Abstract of Haryana (2014–15) reported an average of 323 cars per 1000 people in Gurugram compared to 88 cars per 1000 people in Delhi (Economic Survey of Delhi 2014–15). The city's lack of public transport networks burdens the urban poor in more ways than one. A significant portion of income is spent on commuting to and from work for the urban poor, which could otherwise be spent on addressing other basic needs. At the same time, commuters on foot or bicycle are also at higher risk of road accident injuries, incurring out-of-pocket expenditure on medical expenses. Gurugram's planned expansion of the road network and underdevelopment of public transport has only ensured an ever-increasing demand for cars, which feeds into the demand for even wider roads and rising levels of air pollutants throughout the year. Temperature inversion in the winter months causes heavy smog in Gurugram, affecting visibility for drivers and commuters alike, leading to frequent accidents and greater exposure to harmful air pollutants for people who walk or bicycle as a means of commute. While the city administration has been focused on developing a nonmotorized network plan and bike-sharing facilities, it barely addresses the core issue of negligible public transport with poor feeder services, which could significantly impact public health. It is further established that most real-time pollution sources contributing to Delhi's particle pollution are due to vehicles (CSE 2021). The transportation sector contributes around half of the pollution from solely local sources in Delhi if the total is determined by considering the sources' local contributions alone, excluding those from outside sources.

Without a robust public transport system, city dwellers use individual vehicles to commute; they even use the mass transit system for last-mile connectivity. The identified vulnerability of people due to poor water supply, exposure to dangerous transport, and resultant pollution, along with the absence of sewage connectivity, pose changes to the city's future. About 16 percent of people have no access to safe drinking water, and 27 percent have no access to toilet facilities, as per the 2011 census in India (Census of India 2011). Our earlier study (Das and Choudhary 2024) validated the presence of open defecation and defecation alongside open sewage in Delhi; similar conditions persist in other large cities in India and the Global South. Untreated excreta contaminating groundwater often leads to the consumption of unsafe drinking water, which in turn results in the occurrence of diarrhea, dysentery, parasitic worm infection, and many other water-borne diseases (Ministry of Drinking Water and Sanitation 2011–12; Chaplin 2011; Ortiz-Correa et al. 2016; Patil et al. 2014; Saroj et al. 2020). This leads to a loss of working days, resulting in substantial economic losses (Ministry of Drinking Water and Sanitation 2011–12; Ortiz-Correa et al. 2016). The WASH levels (Water, Sanitation, and Hygiene) remain problematic for a sustainable city and inclusive society in urban India. To develop efficient policy interventions, there is a need for recognition of the existing comprehensive studies related to inequality in WASH, along with its levels and trends (Goli et al. 2011; MoHUPA 2013; Saroj et al. 2020; Sarkar and Choudhary 2020).

6.4 Urban Healthcare System

The prevalence of disease coexists with human existence. The differentiated prevalence of communicable and non-communicable diseases, along with their estimation, has made it possible to relate them to different types of risk and facilitate the analysis of social epidemiology.

Poor preventive measures regarding public health provisions, including clean water and sanitation, mean a wider prevalence of communicable diseases. The disease once appeared needs to be treated to bring back the "healthy human." The concept of health as a fundamental human right for all people, irrespective of "...race, religion, political belief, economic or social condition," was advocated in the WHO constitution (1946).

The novelty of this definition is that, for the first time, it emphasizes "complete physical, mental, and social well-being," contrasting with the previously narrow focus on the "absence of disease or infirmity." This is the most popular and widely accepted definition of health. Last (1988), however, argued this as a "poor operational definition" and presented his definition with measurable attributes for which indicators can be identified and relative vulnerabilities can be mapped for different strata of people. Health is "…a state characterized by anatomic integrity, the ability to perform personally valued family work and community roles, the ability to deal with physical, biological, and social stress, a feeling of well-being, and freedom from the risk to untimely death" (Last 1988). The role performance conception comes close to this definition; however, assessing health in this particular conception becomes impossible when we look at the statistics available from different urban authorities. What is significant to note is that the condition of health in the cities of the global south can still be estimated only based on reported morbidity during large-scale surveys like NFHS (National Family Health Survey), DHS (District Health Survey), or registered morbidity, i.e., hospital-based data.

Health policies worldwide, including India, claim to be directed toward achieving universal health coverage, though the idea of universal and public remains contested. Achieving universal health care requires significant investment in the health infrastructure. However, when we look at the investment, it is less than 1 percent of the gross domestic product (GDP), which is a significant reason for the inadequate health infrastructure in India. On the other hand, due to inadequate public healthcare facilities, people use both formal and nonformal private healthcare facilities, resulting in high out-of-pocket expenditures. Such expenditures push the poor into debt traps and remain indebted to nonformal moneylenders (Choudhary 2010). A health system has to have more than one component, and its efficiency depends upon the integrated functioning of these components to achieve better health care for all. However, in articulation, "all" seems to be inclusive.

In contrast, in operation, "all" is exclusive to those who can afford the ever-increasing high-cost lifestyle and high-end private healthcare services that incur heavy "out-of-pocket" expenditure. A well-functioning health system working in harmony is built on having trained and motivated health workers, a well-maintained infrastructure, and a reliable supply of medicines and technologies backed by adequate funding, robust health plans, and evidence-based policies (Balasubramaniam et al. 2018). It is important to look into the investment pattern in public health care to achieve the ideal "health for all" mission. The rhetoric of "health for all" and the actual implementation of "universal health coverage" seem to be operating on a different scale, as there is a later need for targeted investment in preventive services, including water and sanitation services and accessible public curative health care services. Studies on healthcare expenditure among the urban poor highlight various socioeconomic inequalities in urban areas (Sharma et al. 2020). The social production of health inequality, as shaped by neoliberalism, has to be understood in this historical context of reshaping the capitalist order, primarily based on an unfettered market with the motives of greed and avarice. This divides societies into very rich and very poor groups. A paradox of coexistence of great wealth and great poverty is the reality of the current era, especially in countries like India, which are proliferating. This offends the notion of a just and equal society. Under the pressure of wealthy nations and transnational corporations, the neoliberal reforms implemented by international financial institutions like the International Monetary Fund, the World Bank, and the World Trade Organisation have produced adverse health effects.

The advent of these economic measures in the late 1970s and early 1980s coincided, in fact, with reduced worldwide gains in both economic growth and life expectancy. The question of why health inequalities exist has led to considerable empirical work and much theorizing. The empirical evidence that has been mobilized and studied is vast and goes from survey data to a range of other variables as disparate as crime statistics and voting behavior. Integrating these survey data with psychosocial theories has emerged as one of the most popular and dominant methodological approaches. Indeed, during the 1990s and 2000s, the "psychosocial hypothesis" became widely accepted as conventional wisdom. Wilkinson (2003) and other researchers in Europe and the United States built an original research program on social inequalities in health (Kawachi 2006). These inequalities persist unabated globally, and India is no exception.

6.5 Healthcare Facilities in Delhi

In Delhi, various healthcare facilities are available, both public and private. The Government of Delhi claims to provide accessible and affordable healthcare facilities to the Delhi population and the migratory and floating population. It has been providing primary, secondary, and tertiary healthcare facilities. Certain areas in Delhi do not have access to primary healthcare services within their reach, and these areas mainly comprise the poor and vulnerable population. There has been a wide gap in the demand and supply of healthcare services, evidenced by the long queues in the public healthcare system reported in the media during the COVID-19 health crisis. To fill the gaps in services, the Government of Delhi came up with a four-tier system of healthcare, in which the Aam Aadmi Mohalla Clinics provide primary healthcare.

The neighborhood clinics (aka Mohalla Clinics in Delhi) have been designed to offer accessible, high-quality primary healthcare services within the communities right at their doorstep (Lahariya 2016, 2020; Agrawal et al. 2020). The healthcare crisis is far from over with these initiatives, as private hospitals have gotten land on a subsidized rate with commitments to provide care to the poor; however, they often fail to stand up to their commitments. Research on healthcare facilities for Economically Weaker Sections (EWS) in Delhi offers evidence regarding the percentage of bed occupancy for EWS beds. The bed occupancy data for 2012–13 to 2015–16 shows that the bed occupancy of 41% of the hospitals was below 10%, while only two hospitals had more than 30% occupancy by EWS (Shukla 2018). Even in the Public–Private Partnership (PPP) model, private agencies continue to profit without considering providing healthcare services to the EWS. Many hospitals have been accused of refusing EWS patients, delaying surgery schedules, and charging fees for several procedures. Often, on the flimsy technical ground of an income certificate being "older than six months," treatment is denied. It has been reported that patients have to pay for certain services whenever treatment is provided. Delhi exhibits socioeconomic segregation, and the health-seeking behavior of low-income households differs from that of middle- and high-income households. The public healthcare system, designed to offer healthcare to the less privileged, is scarcely utilized by them for various reasons. The low-income households are forced to use private healthcare facilities, which incur high "out-of-pocket expenditure." Their limited financial capacity burdens them, frequently leading them into a debt trap. It is pertinent that the government should focus on an equitable provision of healthcare to all sections of society, with particular emphasis on primary healthcare services (Gupta and Sengupta 2013). However, the existing reality is that the state has introduced an insurance-based health provisioning system where the poor face multiple discrimination, which sometimes includes paying for treatments that are not required. The land has been provided to various hospitals by different authorities of the Delhi government on lease or at subsidized rates, which came with the clause of serving the EWS patients at no cost.

6.6 Discussion: Future City as Health City

It is known to us that one of the major achievements of human civilizations is the improvement in individual health. Improving life expectancy at birth, reducing maternal and child mortality, and continuing efforts to reduce pain for people at different stages of life have been major achievements. However, the poorer sections of society still spend a sizable proportion of their income on health as catastrophic or out-of-pocket expenditures. Accompanying this is the emergence of what is referred to as the health divide and increasing inequalities in social determinants of health. The social determinants of health, also considered "solid fact" (Wilkinson Marmot 2003), include unemployment, early life, work, food, addiction, transport, and so on. These determinants are further based on access to education, environmental quality, income, preventive activities, and others. Various factors play a role in shaping the health status of individuals and communities in an urban environment. These include the quality of the physical and social environment, stress levels, educational opportunities, the local economy, and urban congestion (Leeuw 2011). The connection between social conditions and health is close to the extent that "the magnitude of health inequalities indicates the impact of social and economic inequalities on people's lives" (Marmot and Allen 2014).

Access to resources, institutional arrangements to address the inability of specific individuals and groups to access resources, and the opportunities to earn a livelihood, which primarily determine income, are dimensions of vulnerabilities that make or break individuals' internal and external resistance to illness. Whether we include indicators on which data is available like socio-demographic variables or proxy variables that indicate individuals' capabilities, persistent inter-urban and intra-urban inequalities are a matter of concern for the future of the city and future cities, which I discuss later in this section. Social vulnerabilities also get expressed in terms of social capital, including social networks and connections, which allow or hinder access to limited public health services in large cities in third-world countries. Two types of indicators, namely the socioeconomic vulnerabilities index and built environmental index, can be used to estimate the level of vulnerability of the masses in large cities, and the intervention in minimizing these vulnerabilities would make us move toward a better and healthier city. The built environmental index (BEI) may include distance to the nearest hospital, population and housing density, age of building stock, average age of sewers and water pipelines, quality municipal roads, exit routes in case of emergency, and so on (Zhong et al. 2022). However, the major challenge in cities in the global south is to have reliable data on these aspects.

Further, the housing stock would matter for those living in the houses, even if they are dilapidated. There is a sizable proportion of people who are either homeless or living as squatters, and without legal rights, the possibility of accessing healthcare becomes more difficult. In such a scenario, universal health coverage is the answer if we plan for a healthy city in the future. The NRHM (National Rural Health Mission), ASHA (Accredited Social Health Activists), and ANM (Auxiliary Nurse) were introduced in 2007 onwards to address the health needs of the masses in India, including vaccination. Delhi introduced the Mohalla Clinic and Mobile Clinic to address the everyday health needs of people at the neighborhood scale. A decentralized healthcare system was introduced in Delhi in 2015 as a form of the neighbor clinics (Mohalla clinics) at the level of primary healthcare centers. However, these clinics are yet to be available across the city and are far from achieving the desired results. More and more people of all social classes, genders, and ages must utilize the services available in these clinics. A specialized form of these clinics for women, i.e., Mahila Mohalla Clinics, is still in its nascent stage and needs to start functioning on a large scale. These specialized clinics are aimed at providing maternal and child healthcare services. These clinics are supposed to be at every K; if this happens, people's everyday health needs can be addressed. If a good network of Mohalla Clinics is established, it definitely has the potential to change the face of Delhi's healthcare system and help reduce the patient burden on the tertiary care public hospitals. Apart from these clinics is a network of primary healthcare centers and Delhi's mother and child care homes. The government needs to take cognizance of this fact and integrate the existing system with the Mohalla Clinics network. The services in these primary healthcare centers should be made more versatile.

Mobile health clinics are an important component of urban healthcare, and their availability and services should be broadened. The health documents of the patients can be stored digitally and need to be accessible at all levels of healthcare facilities so that there is data on ailments, and health planning can be put in place for the future. In absence of the reliable data on registered and reported morbidity, health planning remained only a mirage. Further, if the documents are stored virtually, they can be accessed anytime at any facility level. Full use of digital advancements should be made to make the system integrated. However, the question of privacy and consent adds complexity. Both the central and the state governments claim to offer health insurance. However, in reality, the actual number of patients who benefit from these schemes is questionable. Moreover, insurance-based health services, in contrast to

UHC, are based on collective risk and cater to non-health services, including financial management. There are diseases not covered under the ongoing insurance schemes, like "Arogya Kosh" and "Arogya Nidhi," besides the RSBY and Ayushman Bharat. Universal Health Coverage essentially holds the future of the city significantly. My idea of an anticipatory system that will save the citizens and residents in large cities would be based on universal health care. The experience of East European countries, where UHC is in place compared to out-of-pocket healthcare systems, tells us that it is better for saving lives during a pandemic. Kerala's experience in India has a similar outcome, especially with lower mortality despite high prevalence. The infected persons could be treated and put under quarantine when needed.

Universal Health Coverage (UHC), as defined by the World Health Organization, encompasses the assurance that all people have access to adequate and high-quality health services, including prevention, promotion, treatment, rehabilitation, and palliation, while also ensuring that the utilization of these services does not subject the user to financial hardship (WHO 2023). Draft National Health Policy 2015 acknowledged that the population growth rate has significantly declined and GDP has improved, but health inequity has reversed. The situational analysis leading to National Health Policy 2017 in the opening line reads,

> India is now the world's third-largest economy in terms of its Gross National Income (in PPP terms) with the potential to develop more equitably. As never before, it possesses a sophisticated arsenal of interventions, technologies, and knowledge required to provide health care to its people. However, the inequities in health outcomes continue to remain.

The inherent contradiction in the government's claims persists. Despite asserting the status of the fastest-growing economy and acknowledging existing inequities, governmental investment in India to address glaringly unequal health outcomes based on caste, income, geography, and religion is noticeably lacking. The healthcare needs of the people are changing, and there is a continuous need for the healthcare system to adapt itself. High expectations for a clear roadmap to minimize disparities in access to healthcare services remain unaddressed. The city's future depends on expectation; in other words, we know expectation is the future even though "false consciousness" remains the tool of power to manipulate the masses, and their expectations get shaped in different directions.

Additionally, India was expected to transition toward a "right-based approach to health." However, the flat refusal to entertain the possibility of articulating health as a right, despite soon becoming the third-largest economy, is disconcerting. The existence of health and well-being inequalities across the cityscape is used to clamor for improving living conditions for diverse groups, especially the most vulnerable. Following the broad debate of different SDGs, it is almost agreeable, at least in the literature, that health and sustainability objectives are to be part of urban planning. Pandemic recovery strategies encouraged "temporary urbanism" (Andres et al. 2021) to be considered a solution. They argue that temporary urbanism offers better living conditions and overall well-being and brings alternatives to foster resilience and coping. Temporary urbanism is an anticipatory system; if the temporariness addresses the multiple challenges, it would be a boon for the city's future. The introduction of temporariness in planning and provisioning would, however, be detrimental to the citizens of future cities. The temporary nature of it allows those in power to use it to their advantage, and when there is a possibility of the poor gaining, change the structure. Moreover, the argument is based on the premise that the situation of urban living, that is, the modern city, is given, and we need to develop resilience in the event of any health emergency or otherwise. I wish to argue otherwise, based on the premise that the city of contemporary times needs to be altered fundamentally, at least in two ways: "reduce consumption" and "provide universal health care."

6.7 Conclusions

The world has witnessed over the decade that "fundamentalism," supremacist claims, and dividing the voters are the "universal" means to reach power. In return, power is used to subvert the future by indulging in three precise forms of exploitation: labor, environment, and natural resources. The question is: What future are we looking for? If there exists a "we" or if the world is divided between "us" and "them." On the one hand, hyper-consumerism is the order of the day with technology and innovative measures to spend time and energy enhancing what Singh and Hickel (Singh and Hickel 2023) call capitalogenic diseases and on the other degenerative disease of the poor old persons and diseases of deficiency scar the masses in the same cityscape. The vital choice that humanity has to make is to either engage in a "universalized" echo for the capitalogenic disease or to eradicate vulnerabilities that put millions at risk of being sick. The cities of the future need to vision the long-term intentions, purpose, and aim that will ameliorate the glaring inequities in access to basic services and access to healthcare, meaning, in brief, to provide universal provisioning of basic needs, including water and sanitation services along with ensuring universal healthcare services to everyone.

References

Ameel L (2016) Cities utopian, dystopian, and apocalyptic. The Palgrave Handbook of Literature and the City, pp 785–800

Andres L, Bryson JR, Moawad P (2021) Temporary urbanism as policy alternative to enhance health and well-being in the post-pandemic city. Curr Environ Health Rep 8:167–176

Asaria M, Mazumdar S, Chowdhury S et al (2019) Socioeconomic inequality in life expectancy in India. BMJ Glob Health 4:e001445. https://doi.org/10.1136/bmjgh-2019-001445

Balasubramaniam P, Rao N, Sharma G, Maraganty P, Sathyanarayana TN, Lakew D, Karua F, Bloom G, Ebata A, Gatellier K (2018) Innovations for universal health coverage: a south-south collaboration to transform health systems in Africa and India, innovations for universal health coverage collaboration. Public Health Foundation of India, Delhi

Bhagia M, Bose M (2023) Who owns the city? Neoliberal urbanism and land purchases in Gurgaon, India. Urban Studies. https://doi.org/10.1177/00420980231184784

Bhan G (2009) This is no longer the city I once knew. evictions, the urban poor and the right to the city in millennial Delhi. Environ Urban 21(1):127–142
Björkman L (2015) Pipe politics, contested waters: embedded infrastructures of millennial Mumbai. Duke University Press
Choudhary BK (2005) Waste picking a marginalized activity in urban space. Urban India 25(2):93–116
Choudhary BK (2010) Political economy of waste-pickers: their health and well-being. Trans Inst Indian Geogr 32(1):63–80
Das R, Choudhary BK (2024) Cultural compulsion vis-à-vis poor sanitation in cities: curious case of open defecation in the capital city of Delhi. Indian J Soc Work (forthcoming)
Dupont V (2000) Mobility patterns and economic strategies of houseless people in Old Delhi. Delhi: Urban Space and Human Destinies 99–124
Dutta P, Chaudhuri C (2020) Role of the private sector in escalating medical inflation: evidence from 75th round of NSS data. Econ Pol Wkly 55(22):18–21
Ghertner DA (2014) India's urban revolution: geographies of displacement beyond gentrification. Environ Plan A 46(7):1554–1571. https://doi.org/10.1068/a46288
Goli S, Arokiasamy P, Chattopadhayay A (2011) Living and health conditions of selected cities in India: setting priorities for the National Urban Health Mission. Cities 28(5):461–469
Gupta I, Sengupta P (2013) Health seeking behaviour in urban Delhi: an exploratory study. World Health Population 3(2). https://doi.org/10.12927/whp.17580
Haining K (2021) Mind the gap: Glasgow life expectancy divide. https://the-gist.org/2021/10/mind-the-gap-glasgows-life-expectancy-divide/
Harvey D (2003) The right to the city. Int J Urban Reg Res 27(4). https://doi.org/10.1111/j.0309-1317.2003.00492.x
Lahariya C (2016) Delhi's mohalla clinics: maximising potential. Econ Polit Wkly 15–17
Lahariya C (2020) Access, utilization, perceived quality and satisfaction with health services at Mohalla (Community) Clinics of Delhi, India. J Family Med Prim Care 9:5872–5880
Last JM (1988) Dictionary of epidemiology, 2nd edn. Oxford University Press, New York and Oxford
Latham R, Hicks J (2014) Urban dystopias. The Cambridge companion to the city in literature, pp 163–174. https://doi.org/10.1017/cco9781139235617.014
Leeuw de E (2011) Do healthy cities work? A logic of method for assessing impact and outcome of healthy cities. J Urban Health: Bull New York Acad Med 89(2). https://doi.org/10.10007/s11524-011-9617-y
Ranganathan M (2014) 'Mafias' in the waterscape: urban informality and everyday public authority in Bangalore. Water Alternatives 7(1)
Roy A (2005) Urban informality: toward an epistemology of planning. J Am Plann Assoc 71(2):147–158
Sarkar UD, Choudhary BK (2020) Reconfiguring urban waterscape: water kiosks in Delhi as a new governance model. J Water Sanitation Hyg Dev 10(4):996–1011
Saroj SK, Goli S, Rana MJ, Choudhary BK (2020) Availability, accessibility, and inequalities of water, sanitation, and hygiene (WASH) services in Indian metro cities. Sustain Cities Soc 54:101878
Sassen S (2015) Who own our city and why this urban takeover shall concern us all. https://www.theguardian.com/cities/2015/nov/24/who-owns-our-cities-and-why-this-urban-takeover-should-concern-us-all
Sharma A, Singh AK et al (2020) Out-of-pocket expenditure on healthcare among the urban poor in India. Econ Pol Wkly 55(32):49–55
Shukla S (2018) EWS beds in Delhi: private profiteering on public sufferings. Econ Pol Wkly 53(18):53–59
Singh G, Hickel (2023) Capitalogenic disease: social determinants in focus. BMJ Global Health
Soja EW (2000) Postmetropolis: critical studies of cities and regions. Blackwell

Sulzbach C (2013) From urban nightmares to dream cities: revealing the apocalyptic cityscape. Constructions of space V-place, space and identity in the ancient mediterranean world
Wekesa BW, Styen G (2011) A review of physical and socio-economic characteristics and intervention approaches of informal settlements. Habitat Int 35(2):238–245
WHO (1946) Constitution of the World Health Organization. Geneva
WHO (2023) Universal Health Coverage. Universal health coverage (UHC) (who.int)
Wilkinson RG, Marmot M (eds) (2003) Social determinants of health: the solid facts. World Health Organization
Zhong J, Liu W, Niu B, Lin X, Deng Y (2022) Role of built environments on physical activity and health promotion: a review and policy insights. Front Public Health 10:950348. https://doi.org/10.3389/fpubh.2022.950348

Chapter 7
Anticipating the Future of Iranian Cities: High-Tech Versus Nature-Based Solutions

Mohammadjavad Mahdavinejad, Farzaneh Gharaati, and Mamdouh Ibrahim Zaky Mohamed Galil

Abstract One of the most critical questions about our future is, "What will our cities be like in 2050?". The answer to this question shows our definition of development, humanity, nature, and what we expect from life. The ever-growing world population will intensify the climate change challenge, global warming, etc. Therefore, the research aims to anticipate the possible urban future scenarios and to outline the most prescriptible scenarios for the global south. The research methodology is based on the primary literature of the Discipline of Anticipation (DoA) by Riel Miller, Roberto Poli, and Pierre Rossel for future studies. It was conducted by co-design discussion involving a focus group discussion (FGD) methodology in three phases for expectations and hopes, reframing, and rethinking for flourishing ideas based on future literacy (FL). It was implemented with an interdisciplinary approach ranging from visionary and utopian futures to participatory, critical, or integral futurism by 2050. The conclusion of the research promoted nature-based solutions, socio-economic empowerment, carbon neutrality, high-performance architecture, against irrational ambitious capital-intensive developments. It is to focus on measuring happiness instead of incomes, circular economy instead of short-term gains, nature-based value creation instead of conventional mercantilism, for the MENA region and the global south.

Keywords Discipline of anticipation (DoA) · Global south · Sustainable and green future · Future generations · Low-carbon cities · Carbon neutrality · High-performance architecture · MENA region

M. Mahdavinejad (✉) · M. I. Z. M. Galil
College of Engineering and Architecture, University of Nizwa, Nizwa, Oman
e-mail: mahdavinejad@unizwa.edu.om

M. I. Z. M. Galil
e-mail: mamdouh@unizwa.edu.om

F. Gharaati
Department of Architecture, Tarbiat Modares University, Tehran, Iran
e-mail: farzaneh.gharaati@modares.ac.ir

S. Gonzalez-Arellano and B. Gandlgruber (eds.), *Cities as Anticipatory Systems*, The Urban Book Series, https://doi.org/10.1007/978-3-031-78162-9_7

7.1 Introduction

One of the most crucial questions about our future is, "What will our cities look like in 2050?" This simple question relates to our understanding of our current actions and relationship with nature, which affects our future. On the other hand, the rise in the average surface temperature of the Earth has resulted in adverse consequences linked to climate change on a global scale (Saaty and De Paola 2017). Notably, urbanization is the main factor behind this problem (Khan and Zaman 2018). According to the Department of Economic and Social Affairs of the United Nations Secretariat, the world's population is predicted to increase from 7.6 billion in 2017 to 8.6 billion by 2030 and 9.8 billion by 2050 (DESA 2022). Moreover, it is anticipated that by 2050, approximately 70 percent, or around 6.4 billion people, will live in urban areas (Moriarty and Honnery 2015; Kotzen 2021). While urbanization does offer benefits, such as opportunities for escaping poverty and the potential for long-term sustainability (Khan and Zaman 2018), it also comes with negative consequences. These include pollution, increased resource and energy consumption, and the proliferation of landfills (Saaty and De Paola 2017). As a result, the international community has established various goals to reduce half of global CO_2 emissions, as outlined in the 2016 Paris Agreement (Uwasu et al. 2020). As mentioned, cities are considered the essential drivers of these challenges and are acknowledged as complex and dynamic organisms (Asprone and Manfredi 2015). They offer opportunities and challenges at the same time. This is while, conventional urban planning practices have resulted in excessive greenhouse gas emissions. Moreover, the rise of new needs, globalization, demographic shifts, technological advancements, epidemics, privacy concerns, and data security issues demand the creation of more comprehensive and integrated urban models (Visone 2020) through careful foresight and future visioning in planning and management (Saaty and De Paola 2017). These measures are essential to ensure the ability to adapt to the risks and uncertainties of the future (Likitswat 2019).

Iran, as a country in the Global South, with its millennia-old historical cities, is no exception in this field. While numerous Iranian cities still adhere to traditional planning and models, contemporary Iranian cities have encountered significant shortcomings. Due to these shortcomings, adopting a futuristic vision for sustainable future cities instead of having a nostalgic perspective of the past is imperative. This means a balanced approach that adopts traditional and modern models while reasonably assessing potential future developments is necessary (Sharifi and Murayama 2013).

While Iran's population in 2023 stands at approximately 80 million people, demographic projections for Iran until 2050 anticipate a population of roughly 100 million individuals (PopulationPyramid 2023). Iran is ranked among the top 10 countries regarding carbon emissions, with 75% of its population residing in urban areas. Notably, between 1990 and 2019, Iran's energy consumption increased by 5–8%, a rate approximately five times higher than the global average. A significant portion of this energy consumption occurs in the capital city of Tehran, which has transformed over the past two centuries into one of the world's largest cities, with approximately 9 million residents. Furthermore, buildings, especially in urban areas, contribute significantly to energy consumption, exceeding the global average by 2.5–4 times, with 70% of this consumption stemming from public buildings, which waste up to 60% of their energy (PROVIDAS 2023). Given the demographic growth, cultural changes, and the production of greenhouse gases, particularly in metropolises like Tehran, conducting a thorough study is essential to formulate plans for a greener and more sustainable urban future. As there has not been a comprehensive urban regeneration plan developed for the future of Iranian cities, anticipating and creating scenarios for these cities' green and sustainable future of these cities becomes even more critical. Therefore, this study seeks to anticipate future urban scenarios and outline the most recommended scenarios specifically for the global south, especially Iran.

7.2 Background and Context

According to Wikipedia and UNIDO (2016), a sustainable city, also known as an eco-city or green city, is designed to consider its social, economic, and environmental impacts, often referred to as the three pillars. A sustainable city is a platform for low-carbon industrialization, turning today's cities into livable economic powerhouses with clean energy innovations. Indeed, it aims to provide a resilient habitat for the present generation without compromising the ability of future generations to have the same experience. Sustainable cities are dedicated to minimizing required energy, water, and food inputs and drastically reducing waste, output of heat, and air pollution—CO_2, methane, and water pollution. The significance of this endeavor stems from the fact that cities significantly influence the lifestyles of future generations (Likitswat 2019; Nakagawa 2020).

However, research findings indicate that among the three dimensions—social, economic, and environmental—environmental indicators have received less attention in proposed future city frameworks (Khan and Zaman 2018). The importance of these indicators cannot be overstated, as neglecting them could lead to cities becoming unlivable due to the impacts of climate change, necessitating the construction of entirely new cities (Moriarty and Honnery 2015). Drawing from the insights of authoritative articles and a comprehensive review of ideas and concepts, it is evident that creating sustainable cities for the future demands a multidisciplinary approach and comprehensive management through co-design (Riffat et al. 2016).

Over the years, various terms and labels have been employed to envision "ideal" and "future" cities. In 1516, Thomas More introduced the concept of Utopia, which depicted an idealized, fictitious island nation characterized by a lack of private property and a high degree of religious tolerance (Lang 2007). Utopia is a term that conceptually relates to future cities and is usually interpreted as an ideal city or society. Since both strive for more favorable and better living situations through innovation, utopia concepts can inspire future cities. Both serve the areas of technology, environmental sustainability, harmony with nature, justice, and social equality (Bellagamba 2016).

Different names and designations have been assigned to alternative visions of future cities over time:

- Age-Friendly City
- Compact City
- Creative City
- Circular City
- Eco-City
- Eco-Smart City
- Garden City
- Global City
- Green City
- Hanging City
- Livable City
- Low-to-Zero Carbon City
- Regenerative City
- Resilient City
- Sharing City
- Smart City
- Vertical City
- Zero Waste City

For instance, a *compact city* represents a densely populated urban area with reduced parking spaces and a strong emphasis on public transportation. Typically, compact cities feature limited population density and abundant green spaces and recreational areas, which can contribute to a more sustainable lifestyle for future generations (Saaty and De Paola 2017). Conversely, *Vertical Cities* constitute a network of interconnected, eco-friendly, self-sustaining towers that extend up to a mile into the sky, reaching up to one kilometer or a mile (Kotzen 2021). The concept of a *Hanging City* is conceived to preserve natural landscapes and prioritizes an ecosystem-based design, along with achieving energy self-sufficiency. Meanwhile, *Green Cities* are designed by harnessing natural forms and resources to construct future residences (Saaty and De Paola 2017). Furthermore, several popular ideas, such as Circular Cities, Smart Cities, Eco-Cities, and Eco-Smart Cities, have gained traction in the discourse on future urban development.

Circular cities represent a novel and practical approach to addressing the intricate and pressing challenges associated with sustainable urban development, arising in response to the dynamics of urbanization. The notion of a "circular city" is derived from the principles of the circular economy model. Circular city strategies align closely with the objectives outlined in the Sustainable Development Goals (SDGs), particularly emphasizing SDG11 and SDG12 (Gravagnuolo et al. 2019).

Ecology, as defined, involves the scientific examination of the interplay between organisms and their surroundings. Consequently, one can infer that ecological cities represent intricate and manufactured ecosystems with a strong synergy between society, economy, culture, and nature. These *ecological cities* are fundamentally concerned with recycling materials, achieving efficient energy utilization, implementing regulatory mechanisms, maximizing economic advantages, fostering societal harmony, and advancing the coexistence of humans and nature (Li and Yang 2015).

Smart city originally emerged as a simple reference to delivering energy and technology solutions to pressing sustainability issues, such as climate change and resource scarcity (Shah 2023). Over time, this concept has evolved and acquired diverse interpretations (Visone 2020). It includes comprehensive plans encompassing various urban infrastructures and functions, including services, commerce, transportation, communication, water, and energy (Saaty and De Paola 2017; Javed et al. 2022).

In contemporary discussions, the most favored concept is that of an *eco-smart city*—a city that focuses on economic, social, and spatial aspects to prioritize environmental protection equally. It emphasizes preserving the environment, conserving natural resources, limiting carbon dioxide emissions, reducing water consumption, and utilizing renewable resources. This concept emerges as a response to numerous factors, such as climate change, the vulnerability of the global economic system, urbanization, the digital transformation of society, higher expectations of life quality, demographic changes, and a growing public awareness of the need to protect the environment for the benefit of current and future generations. It represents the future's form—a shift from an industrial to a post-industrial civilization (Cheng and Xie 2011) (Fig. 7.1).

Fig. 7.1 A vision of an eco-smart city (Joachim 2018)

7.3 Methodology

Future thinking is an age-old human phenomenon. However, modern futures research advances far beyond that, and the systematic exploration of alternative futures is defined as foresight instead of prediction or prophesizing. Modern futures research was established in the late 1940s as a scientific field (Heinonen et al. 2017). All attempts to "know the future" mean thinking and anticipating that the future is equally present in all phenomena, conscious or unconscious, physical or ideal. The complete prediction of change in our world is practically and theoretically unattainable and nothing more than a virtual existence (Pillkahn 2008).

Regarding prediction or anticipation, "a predictive system is a system that contains a predictive model of itself and its environment, which allows this system to change its state at one moment by the model's predictions about a later moment." Practically speaking, the problem is that the data and models used to anticipate the future are incomplete and need to be supplemented. Theoretically, the challenge of anticipation is that our world is in some sense "creative" and that innovation happens. Many forecasting exercises are based on a modeling approach tested using past data (Poli 2017). Another fundamental challenge when imagining the future is the absence of future generations in the process, which causes intergenerational conflicts (Hara et al. 2021; Uwasu et al. 2020; Timilsina et al. 2020). That **is while creating** a series of failures, such as global warming and biodiversity loss, that will affect future generations (Saijo 2020). Notably, the future is unpredictable, and we can only create images and ideas of events we will face. Also, there is not one future but multiple possibilities. Furthermore, the future can be influenced based on decisions (Heinonen et al. 2017; Pillkahn 2008). Moreover, as we move further from the present, the number of future possibilities increases because various other factors influence what will happen (Pillkahn 2008) (Fig. 7.2) (Many attractive scientific researches have been carried about the Scenario Cone, see: Gall et al. 2022).

Before conducting anticipation research, defining the specific knowledge area and choosing the approach strategies is crucial. Additionally, it is logical to gather some references or real-world examples to provide context. Evaluating anticipations poses a challenge because the future does not come with concrete data, making it hard to determine the success of predictions. Even if predictions eventually align with reality, there is often no real-time assessment method (Miller et al. 2013). We can turn to Futures Literacy (FL) to tackle these challenges. Like learning a language, FL equips us with the skills to navigate the future effectively. It is like learning to read and write in a new language of possibilities (Miller 2018). FL involves understanding "what to know," "when to know," and "why it is important to know." One of its strengths is its ability to distinguish between different ways of using future knowledge (Miller et al. 2013).

Anticipation has three general levels:

- The first level (forecasting) is part of the correct prediction of future studies. Its models tend to consider a very short or very long time horizon.

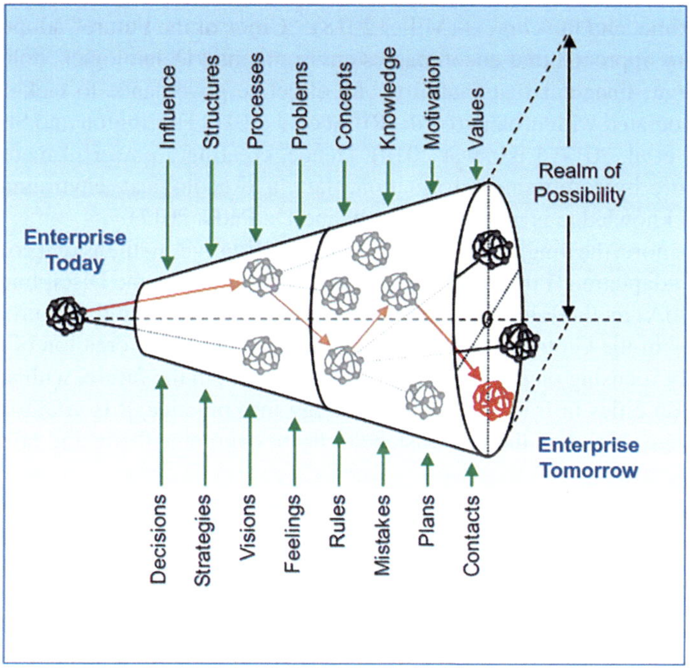

Fig. 7.2 The scenario cone: the realm of possibility (Pillkahn 2008)

- The second level (foresight) includes most traditional futures studies. It is not predictive, so the main output of most foresight exercises is the production of several possible futures. Foresight is often qualitative and includes and even focuses on discontinuities. As practiced, foresight usually shows a limited acceptance of complexity.
- The third level (anticipation) is based on the results obtained from forecast and foresight models, and the goal is to implement them in decision-making and actions.

Anticipation exhibits many temporal patterns, from micro anticipations embedded in perception to usually longer forms of social anticipation, ranging from seconds to years and decades (Poli 2017).

During the startup phase, several critical questions appear: "Who should be invited to take part?", "how should they be extended invitations?" and "Where should the process take place?" (Miller 2018). The challenge here is that participants often need to shift their perceptions of the future to shape a shared vision collectively. This typically involves collaborative work within small groups. Through this process, participants often appreciate that the future comprises diverse aspects, both pessimistic and optimistic. It is important to note that the choices regarding participants and settings depend on factors such as the specific group, the topic, available time, and the context (Miller 2018). "Cities of the Future" adopts an interdisciplinary approach that encompasses environmental technologies, holistic urban development, financial sustainability, and effective governance to tackle the challenges associated with urban growth (Riffat et al. 2016; Harrington and Stryja 2021; Heinonen et al. 2017; Likitswat 2019). Hence, creating a powerful meta-narrative of the future by combining information, data, and ecological, environmental, and individual knowledge is necessary (Saaty and De Paola 2017).

Furthermore, the magic concept of a "shared vision" strengthens real commitment instead of adaptation (Pillkahn 2008). Therefore, in applying the Discipline of Anticipation (DoA) methodology in future urban studies, developing future urban visions for regions in the Global South is imperative. This needs the creation of scenarios, specifically focusing on the emergence of green cities in the future, with a particular emphasis on cities in Iran. To put this strategy into practice, it is vital to recognize the challenges based on the circumstances, try to overcome them, and offer answers for each.

7.4 Vision for Urban Future of the Global South

Global South countries are primarily situated in the Southern Hemisphere, including regions such as Asia (excluding Japan, Taiwan, Singapore, South Korea, and Macau), Central America, South America, Mexico, Africa, and the Middle East (excluding Israel) (Cruz 2015). In the future, cities in the Global South are anticipated to face rapid urbanization while concurrently experiencing rising temperatures due to the impacts of climate change and urban development (Bardhan et al. 2020). Many articles propose the implementation of "smart city" technologies as a remedy for urban challenges in the Global South (Borkowska and Osborne 2018; Visone 2020). However, it is essential to recognize that relying solely on these technologies will not be sufficient for shaping and achieving a green and sustainable future. This issue requires careful management and the importance of continuous study for planning future cities. In addition to exploring the ideas and concepts, there is a need to consider solutions based on nature-based solutions approaches and advanced technology.

In urban development, future cities share a common aspiration, regardless of the labels attributed to them: to furnish their citizens with the most optimal and desirable urban experiences (Khan and Zaman 2018). Furthermore, involving local communities and civil society in making decisions about urban matters is essential. It is crucial to listen to the opinions of all people during the urban development process (Saaty and De Paola 2017). Our investigation into the concept of ideal cities has led us to identify two conceptual paradigms: high-technology and tall buildings approaches and nature-based solutions.

7.4.1 High-Technology and Tall Buildings

Tall buildings will be essential in future cities because of the world's rapidly increasing urban population and the migration of people from rural-to-urban areas due to growing security concerns. Tall buildings have critical features like preserving agricultural land and offering multi-level connectivity, height, and multifunctional and biophilic content (Ali and Al-Kodmany 2012). They stand out for their ability to connect multiple towers with skywalks instead of isolated single towers. While most people imagine the vertical city in the air, it is conceivable that these cities could easily be extruded underground as well, especially since some functions do not require natural light (Fig. 7.3). This concept of the "vertical city" creates a network of three-dimensional green spaces throughout the city, unlike traditional cities with separate parks and gardens, the biophilic vertical city creates a network of three-dimensional green spaces throughout the city (Kotzen 2021). Technological advancements make all these innovations possible (Riffat et al. 2016).

7.4.2 Nature-Based Solutions

Nature-based solutions (NbS) emerged in the late 2000s, initially introduced by the World Bank and IUCN (2009), primarily emphasizing the significance of biodiversity preservation and mitigating climate change. The European Commission has recently defined NbS as "actions inspired, supported, or imitated from nature" (Kabisch et al. 2017). These actions involve the creation of urban green spaces, such as parks and street trees, which can help alleviate high temperatures in cities and regulate air and water flow. Architectural solutions, including green roofs and wall installations, can also reduce temperatures and conserve energy (Kabisch et al. 2017). NbS aims to develop sustainable urbanization to improve economic growth, enhance environmental conditions, restore degraded ecosystems, adapt to climate change, implement carbon reduction measures, improve environmental risk management, and promote multifunctionality (Atanasova et al. 2021).

The concept of NbS is intricately linked to other ideas such as sustainability, resilience, ecosystem services, human–environment interaction, reinforcement, reconstruction, co-creation, and co-design of urban green networks, as well as green/blue infrastructure and biomimicry concepts. Compared to conventional approaches, NbS presents a more efficient and cost-effective development strategy (Lafortezza et al. 2018; Dushkova and Haase 2020). Concepts like ecosystem-based adaptation (EbA) (focused on providing services that support human adaptation to climate change), green infrastructure (GI) (aimed at supporting native species), and ecosystem services (ESS) (aimed at facilitating people's benefits from ecosystems) are closely intertwined with the NbS concept within scientific literature (Kabisch et al. 2017). A notable characteristic of NbS lies in its broad nature and scale, ranging from expanding forested areas to absorbing gaseous pollutants and spanning from micro to

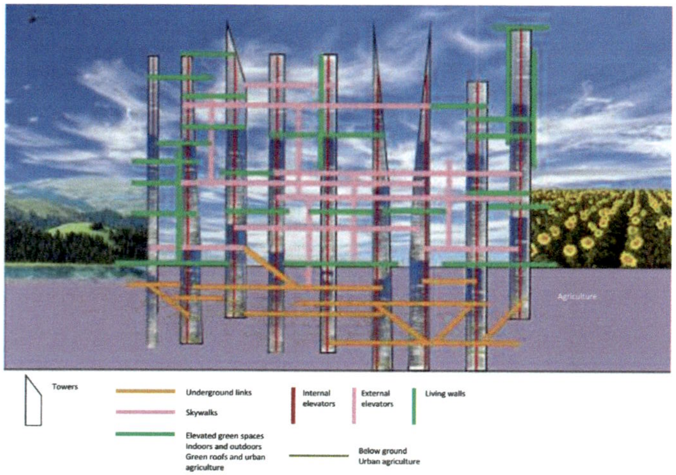

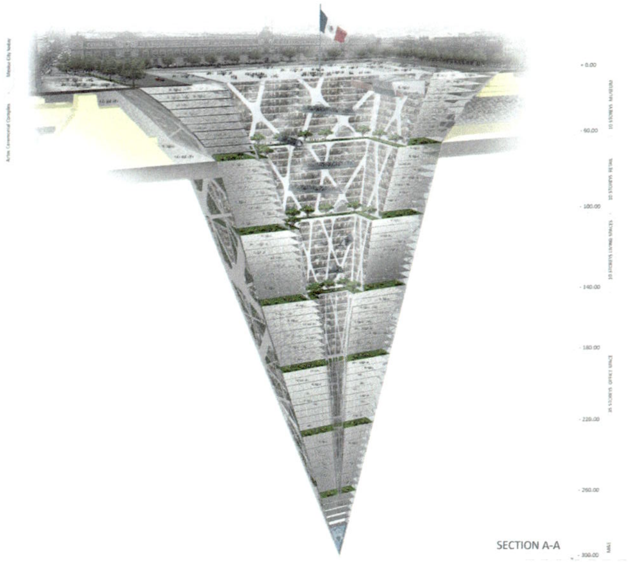

Fig. 7.3 Left: concept of the biophilic vertical city, right: Mexico city's proposed 75-storey underground buildings in the future (Kotzen 2021)

macro scales. This encourages collaborative approaches for co-design, co-creation, and co-management, often called "co-co-co" (Kabisch et al. 2017; Lafortezza et al. 2018). The concept of a circular economy, which is also connected to nature-based solutions, holds significant importance. Circular economy (CE) is a system that aids cities in accomplishing a transition from managing their resources in a linear to a circular manner. At its core, fundamentally, the primary challenges of circular economy (CE) are resource conservation and reuse, waste reduction, and resource utilization cycle establishment (Atanasova et al. 2021; Langergraber et al. 2021). Key aspects of CE include the recovery and preservation of the water cycle (including the management of rainwater), water treatment, and waste treatment, alongside efforts in recycling and reusing resources. Simultaneously, CE strongly emphasizes enhancing energy efficiency and harnessing energy recovery methods (Atanasova et al. 2021). Furthermore, it highlights the importance of building reuse, epitomized by the concept of Zero Energy Buildings (nZEB) (Riffat et al. 2016). Embracing vertical greening systems and green roofs, promoting food and biomass production, and facilitating the creation of (public) green spaces also rank among CE's fundamental principles (Langergraber et al. 2021; Heinonen et al. 2017). These principles necessitate a holistic, cross-sectoral, and interdisciplinary approach (Atanasova et al. 2021).

7.5 Scenarios for the Green Future of Iranian Cities

7.5.1 Traditional Cities of Iran

Historical urban planning in Iran can be categorized into two primary periods: pre-Islamic and post-Islamic. Our understanding of pre-Islamic cities primarily relies on archaeological findings, and as a result, access to direct documentation about the social life in these cities is scarce. Consequently, figuring out specific urban patterns for this period is challenging. In the post-Islamic era, a distinctive transformation occurred in urban planning as new urban forms emerged. The combination of pre-Islamic and Islamic concepts gave these post-Islamic cities a unique identity that distinguished them from their pre-Islamic counterparts. To gain a deeper understanding of cities in Iran, it is essential to familiarize with the main urban elements and their functions:

- *Mosques*: After the Arab invasion, structural changes were made in cities to meet Islamic needs. The mosque was one of the main elements of a square connected with other elements.
- *Bazaar*: Is originally developed to perform commercial and production functions and usually has a linear shape.
- *Meydan*: It was created in the pre-Islam era.

- *Mahala*: Traditional Iranian cities were divided into several residential neighborhoods called Mahala. People with similar ethnic and religious backgrounds were gathered together in the same neighborhood. Narrow winding alleys connected the center of the neighborhood with the market on one side and ended in a dead end on the other. The alleys had a hierarchy and became narrower away from the *Bazaar*. These winding passages were covered intermittently to provide greater security against horse raiders if the walls were breached.
- *Zorkhaneh* (house of power): It was created in the pre-Islam era. They were not only places for physical exercise but have been centers of social and political activity for centuries.
- *Hammam* (public baths): They were health and social places where people go to bathe, massage, shave, and talk, and they can be found in all the traditional cities of the Islamic world.
- *Persian garden*: It is an exceptional and even unique evidence of cultural traditions and has become a prominent feature in private residences, palaces, public buildings, and collections related to charitable or religious institutions, such as tombs (Sharifi and Murayama 2013).

Therefore, the reasons for the geometric formation of traditional Iranian cities can be summarized as climate, agriculture, and religion. In terms of climate, for example, the streets of Tehran are arranged in the main directions, and the houses face north to take advantage of the summer breezes from the Alborz mountains. From a religious point of view, because Muslims must face Mecca when they pray, Iranian mosques, the main element of Iranian cities after Islam, are rectangular with open courtyards, and the Qibla wall must face southwest and toward Mecca. The streets are often parallel to the mosques, and the grid system may have been developed along the streets around the mosques perpendicularly and toward the southwest (Bonine 1979).

7.5.2 Contemporary Cities in Iran

Contemporary cities in Iran have experienced significant transformations over time. While the initial steps toward modernization began during the period of Naser al-Din Shah in the 1870s, with his visit to Paris serving as a catalyst, the peak of these transformations occurred in the early 1920s when Iranian cities witnessed substantial structural changes. Unfortunately, urban renewal principles were often applied without considering historical growth patterns or preserving local cultural values. The introduction of automobiles marked a significant moment in the urban landscape, leading to street systems designed primarily to accommodate motorized transportation needs. The Approval of the Law on Municipalities in 1930 and the Street Widening Law of 1933 played crucial roles in this transformation, as straight and axial street networks began to disrupt the traditional urban fabrics.

Consequently, the cohesive urban structure started to erode, causing the fragmentation of key elements of the traditional city. Over time, these elements lost their functionality and efficiency, necessitating a redefinition. Historically, urban routes were meticulously designed to ensure pedestrians' comfort and give precedence to environmental considerations. However, in contemporary developments, streetscapes have shifted their orientation toward cars, often neglecting pedestrian safety. Traditional cities were adept at creating outdoor spaces that provided relative respite from adverse weather conditions. Still, new developments often disregard these principles, resulting in uniform dimensions across the country, regardless of the local context.

Additionally, there is insufficient consideration of hierarchy in the new street patterns, diminishing the traditional order from public to private spaces. As activities are segregated in modern zoning systems, residents spend less time in their neighborhoods, impacting social interactions and reducing the sense of belonging to these areas. Beyond physical changes, increased rural-to-urban migration and the Iran-Iraq war further complicated the urban landscape. The war forced residents from war-torn regions to migrate to other cities in search of land to build their homes under the Urban Land Law of 1987. These multifaceted factors have collectively shaped the contemporary cities of Iran, highlighting the need for a holistic understanding of their development and transformation (Sharifi and Murayama 2013) (Fig. 7.4.) Hence, exploring city streets is crucial for shaping future city concepts, especially in Iran, where significant street changes followed modernization. As Boudreau (2022) argues, studying streets can reveal hidden ways to influence the future.

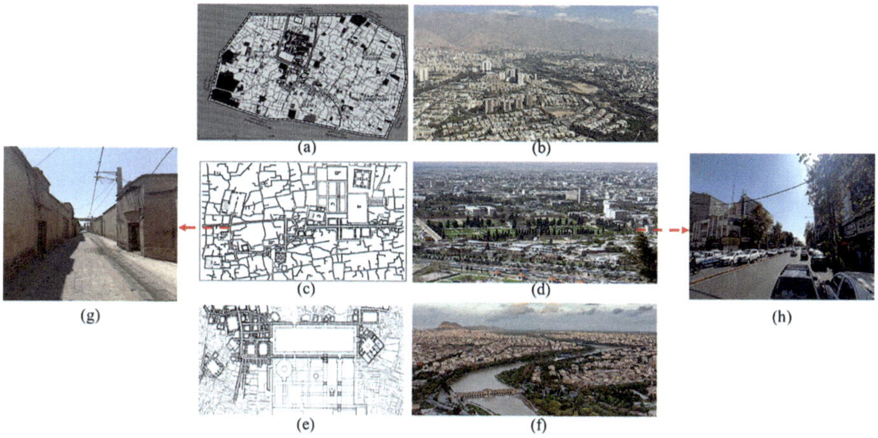

Fig. 7.4 **a**, **c**, **e** Historical context of Iranian cities: Tehran, Shiraz, and Isfahan, respectively, **b**, **d**, **f** The current view of the Iranian cities: Tehran, Shiraz, and Isfahan, respectively (Jamei et al. 2021, Shirazi and Falahat 2015), **g** Alleys and historical districts of Shiraz city, **h** The current streets of Shiraz city (authors 2023)

7.5.3 Presenting Scenarios

Through an examination of ideal city concepts, we have outlined two distinct scenarios that envision the future cities of Iran in 2050. These scenarios are integral to the evolution of urban landscapes, offering unique solutions to meet the challenges of sustainable urbanization. We present these scenarios as two compelling paths for urban development in the coming decades, focusing on the most populous cities in Iran, including Tehran, Isfahan, Shiraz, Mashhad, Tabriz, and Karaj. The finer details of these scenarios will be left to urban planners' and experts' expertise and contributions to avoid any orientation. With the DoA method, we are operating on a distinct time frame. We decided on the year 2050, around thirty years from now.

Scenario 1: Technology and Tall Buildings

The first scenario centers around integrating technology and tall buildings, resulting in intelligent and vertical cities. This vision hinges on making intelligent various combinations of urban life, people, the environment, and lifestyles (Visone 2020). Additionally, it entails the construction of interconnected tall buildings that redefine urban living. This approach stimulates sustainability through innovative technologies and vertical urbanization.

Scenario 2: Nature-Based Solutions

The second scenario promotes nature-based solutions, combining the principles of an eco-city, a green city, and a zero-carbon city. This vision considers city regeneration, ecosystem restoration, and the sustainable utilization of resources and energy (Lafortezza et al. 2018). It incorporates green/blue infrastructure into contemporary urban planning (Gearey et al. 2019).

These two scenarios have been chosen based on their capacity to enhance sustainable urbanization and address critical urban challenges. Both nature- and technology-based solutions contribute significantly to the following aspects:

- *Ecosystem Restoration*: The scenarios support ecological health by making restoring damaged ecosystems and vital services easier. This is crucial for densely populated Iranian cities like Tehran, which face pollution challenges.
- *Climate Change Adaptation and Mitigation*: These scenarios integrate climate change strategies by reimagining gray infrastructure as part of green /blue infrastructure. Because it considers the potential of Iranian cities to adopt green/blue infrastructure, leveraging their advantageous locations.
- *Risk Management and Resilience*: They enhance risk management and urban resilience by offering multifunctional nature-based designs, reducing pollution, protecting biodiversity, mitigating heat stress, and enhancing water retention (Dushkova and Haase 2020). Most Iranian cities like Tehran rank among the cities with significant carbon emissions.
- *Microclimate Regulation*: They regulate microclimates, prevent floods, purify water, and promote sustainable food production (Langergraber et al. 2020). Given

their access to rainwater during various seasons, Iranian cities possess substantial potential for harnessing microclimates to obtain clean water.

- *Circular Economy*: These scenarios, especially Nbs, encourage long-term material design, maintenance, repair, reuse, reconstruction, renovation, and recycling, reducing waste and minimizing environmental impact (Langergraber et al. 2020). This is especially relevant for many Iranian cities with rich historical heritage, as reusing resources aligns with the circular economy concept. Additionally, buildings that release greenhouse gasses and carbon emissions may be able to mitigate their effects through circular renovation.
- *Job Creation:* These scenarios generate new employment opportunities that align better with the needs of urban populations (May et al. 2022). Given the substantial population, job creation holds significant importance.

7.5.4 Results

A series of expert meetings, spanning three sessions of approximately one and a half hours each, were held to deliberate and strategize on the future cities of Iran in 2050. The experts were generally chosen based on their experience and knowledge background in the context of Iranian cities. The objective was to apply their knowledge and insights to imagine the urban landscape of Iran's future based on considered scenarios and the four distinct images that depict various trajectories:

1. Continuous Growth: Illustrating a future that aligns with conventional expectations.
2. Transformation: Envisioning a future that exceeds expectations, striving for substantial improvement.
3. Collapse: Picturing a future worse than expected, marked by challenges and failures.
4. Protective/Disciplined Society: Imagining a future that unexpectedly departs from expectations (Cruz 2015; Boyko et al. 2012).

Eight scenarios appeared after determining the four trajectories and considering them with the two basic scenarios. During the initial meeting, experts discussed and decided that the most promising approaches for shaping Iran's future cities are the "Transformation" and "Protective/Disciplined Society," which correspond to numbers 2 and 4, respectively. This decision was made with the common goal of collaborating toward improving Iran's future. Nevertheless, it was acknowledged that substantial challenges and obstacles must be anticipated and addressed to realize these scenarios successfully.

During the following discussions, a crucial realization occurred when specialists acknowledged that depending only on natural or technological solutions was not enough. They concluded that the cities of Iran in the future of 2050 would require a well-balanced combination of all strategies and scenarios. However, they chose to place the scenarios on the "Transformation" mode and make judgments in this mode because the "Protective/Disciplined Society" mode is not very certain in Iran until 2050, based on accurate data.

In the third session, the experts were asked to imagine themselves as the Imaginary Future Generation (IFG) facing an intergeneration dilemma (Hiromitsu et al. 2022). It was marked by challenging and controversial discussions as experts collaboratively presented various solutions and strategies tailored to the unique characteristics of Iran's future cities in 2050. These deliberations centered predominantly on densely populated cities, including Tehran, Isfahan, Shiraz, Tabriz, Mashhad, and Karaj.

Through these rigorous and insightful expert meetings, a comprehensive framework for the future cities of Iran in 2050 began to take shape, emphasizing transformation according to a harmonious blend of nature-based and technology-driven solutions.

We also faced several practical challenges: (1) creating scenarios; (2) selecting participants; (3) forecasting uncertainties; and (4) estimation of effective variables until 2050. In order to address the first challenge, the statistical data of Iran in the last ten years and articles on predicting Iran's future were studied. However, we refrained from providing specific statistics and data to the participants as we opposed providing them with any orientation. We tried to select urban development and planning experts for the second challenge because this type of meeting is the first of its kind in Iran. In order to evaluate the third challenge, we consulted other experts on the concepts. The fourth challenge was the hardest one. The research results show that climate change and global warming are the most obvious factors affecting Iran's future. However, this method can be combined with other quantitative methods under the condition of completing and solving the challenges in future research.

7.5.5 Discussion

The experts' conceptualizations of the scenarios to represent Iran's future cities in 2050 are provided in this section. Besides, a comparison with the findings of our studies is made. Thus, challenges about the scenarios will be discussed depending on the methodology.

Scenario 1: Technology and Tall Buildings with Transformation Mode

Urban areas in the global south, especially Iran, grow gradually rather than all at once. In this scenario, the conceptual framework of sustainable cities in all social, economic, and environmental dimensions is used to imagine the sustainability and intelligence of cities. Technology is essential in every aspect; ideas and solutions can only be implemented with a fundamental shift (Wickenberg et al. 2021). The historical context, dominating beliefs, customs, and past behaviors reflect the identity of Iran's traditional cities, such as Tehran, Isfahan, and Shiraz. As a result, there is a high potential for value creation through technological integration in historical contexts and legacy. Furthermore, it is possible to apply traditional patterns in various contexts and cities, focusing on the social aspects of evolving and developing technologies.

High-rise structures, a new generation of buildings, referred to as "green skyscrapers/earth-scrapers" or "zero energy buildings," are becoming more energy efficient and capable of producing as much energy as they consume. Furthermore, vertical farms have a bright future. The digital revolution and technological advancements in construction materials are expected to influence the architectural style and functions of buildings in Iranian cities (Mahdavinejad et al. 2024). This strategy needs the participation of stakeholders, designers, and policymakers in collaborative processes that will produce a city model that intelligently manages its resources, encourages the development of innovative urban transportation systems, and strives for economic sustainability and energy efficiency. On the other hand, quality of life considerations in this scenario go beyond innovative information and communication technologies and reach urban sustainability, which is achieved through optimizing resources, increasing energy efficiency in buildings, and reducing land use. Finally, it follows the precise and comprehensive mechanisms of a smart city (Wickenberg et al. 2021) following intelligent optimization methods (Fakhr et al. 2023; Shirzadnia et al. 2023; Najafi et al. 2023; Goharian et al. 2023).

The challenge we may face in this scenario is the construction of tall buildings; while they symbolize modernization and progress, they create significant demands on infrastructure and transportation systems. In addition, they can have a profound effect on the historical contexts and skyline of the city. However, a potential future problem is an unhealthy competition between designers and developers who are only focused on achieving greater prestige, potentially ignoring citizens' real needs and preferences. Furthermore, there is a danger that local culture, ecology, and contextual relevance may be overlooked in the pursuit of architectural grandeur. This idea is probably because the experts are well aware of the events of the development of contemporary cities in the 1900s for Iranian traditional cities, and they hope to prevent it from happening again.

Scenario 2: Nature-Based Solutions with Transformation Mode

According to experts, since most of Iran's future challenges are environmental, climatic, and rising temperatures, presenting this scenario is very important for Iran. To illustrate this scenario, the experts highlighted some examples of passive methods of renewable energy usage in traditional Iranian cities (Dezfuli et al. 2023; Saadatjoo et al. 2023). Using traditional badgirs (wind deflectors), Asbad (windmills), shading in narrow alleys, and geometric proportions of elements (Dezfuli et al. 2024; Latifi et al. 2023), especially in hot and dry cities are a few examples of these techniques. Using these traditional patterns not only helps to conserve energy and improve energy efficiency in buildings and cities but also contributes to value creation. These views are most likely based on the expert's knowledge in the field of Iranian cities, especially traditional ones. Like traditional cities, the residents' biological, social, and perceptual characteristics and prevailing weather conditions must be considered in contemporary cities. These factors play a significant role in evaluating the happiness of citizens. When adopting nature-based solutions, aesthetic appeal and values are essential to citizens and require collective governance. This approach can seamlessly integrate into various urban programs (Frantzeskaki 2019).

Another idea that experts mentioned was the circular economy, which leads to clean production methods by reusing scraps and waste. Iranian cities have a high potential for this idea because the historical context and traditional patterns are highly flexible (Cavaleiro de Ferreira and Fuso-Nerini 2019; Gharaati et al. 2023). This transformational change can promote sustainable growth, stimulate economic development, and reduce the adverse effects of climate change while minimizing the consumption of raw materials (Lakatos et al. 2021). This is probably because experts seek to introduce innovative forms of value creation and break away from traditional ways of value creation. They aspire to harness the boundless potential of creativity and innovation to cultivate unique forms of value that resonate with the evolving dynamics of Iran's urban landscape.

Challenges and Considerations

The proposed scenarios, although promising, come with a set of major challenges that must be addressed holistically. The most critical challenges are legal deficiencies, lack of capacities, and public awareness in Iran. Both technology-based and nature-based approaches face these common challenges. Experts also mentioned the concern about over-reliance on technology. While technological advances offer significant benefits, over-reliance on them can raise questions about the flexibility and sustainability of these cities.

On the other hand, when it comes to Iranian cities, there is a need for cultural preparation of Iranian cities to accept the proposed scenarios. In addition, considering factors such as climate, economy, and the unique historical context of many Iranian cities, special solutions must be devised. Another challenge is the relationship between tall buildings and considerations related to the aesthetics, culture, and overall texture of Iranian cities. Determining how to address this issue is critical because implementing and planning these solutions requires careful study. Finally, applying advanced technology approaches in Iranian cities creates a challenge because it requires significant structural and fundamental changes. In addition, a significant challenge is the potential for high real estate prices that usually appear very quickly in Iranian cities. As Iranian cities embrace intelligent and sustainable features, property values may increase, potentially making housing unaffordable for certain population segments.

In addition, the complex communication between Iranian intelligent cities and their non-smart counterparts creates challenges. Ensuring the integration and seamless cooperation between these two types of cities requires careful planning and implementation. High electricity costs are another concern, especially in the context of high-tech urban infrastructure. Imagine a city intricately connected and controlled through satellite and Internet networks, highlighting the advantages and disadvantages of such connectivity. While offering greater security and convenience, it raises legitimate concerns about personal space and privacy. Maintaining privacy in Iranian cities is a significant concern, which makes it an important assignment. On the other hand, the potential vulnerability of the entire system to the Internet, such as disruption, is a significant concern. Furthermore, the advent of new technologies may inadvertently lead to a decrease in physical activity in individuals, potentially leading to long-term health problems such as increased rates of disease and obesity (Shichiyakh et al. 2016).

Therefore, instead of relying on conventional measures such as income and short-term economic gains, our research addresses a new paradigm that places city dwellers' happiness and well-being at its core. Experts emphasize the new urban vision, which makes the happiness and well-being of Iranian city dwellers its primary goal. They support a shift toward comprehensive urban well-being and emphasize metrics that measure the city dwellers' quality of life, satisfaction, and happiness rather than income. This change also emphasizes deviating from the conventional pursuit of immediate financial gains. Therefore, instead of income, experts suggest a circular economy, instead of short-term profit, value creation in new forms instead of traditional forms.

7.6 Conclusion

This study aims to provide new strategic concepts and scenarios for the future cities of the global south in 2050, with a focus on Iran. To achieve this, a series of meetings were conducted with five experts with deep knowledge of traditional Iranian cities based on the Discipline of Anticipation (DoA) and Future Design (FD). A range of alternatives can be described for Iran's future cities in 2050 through scenario-based literature. This makes it possible to compare future differences directly. However, the outcome of the experts' discussion brought two scenarios: technology and tall buildings and nature-based solutions with the transformation mode in 2050.

According to the findings, considering both scenarios simultaneously is more beneficial than doing so. Furthermore, the quality of life and the demands of future generations are intrinsically related to their happiness. Given that both scenarios have a high chance of being welcomed in Iranian cities, to overcome environmental challenges, details, and solutions of the scenarios should be designed under Iran's traditional and cultural values and the design principles of Iran's historical cities. It is also necessary to use each city's specific characteristics, such as hierarchy and street orientation, to consider future climate changes.

In characterizing the capacities to anticipate futures in Iran, two scenarios show comprehensive potential. In the first scenario, Iran has promising developments in technology and infrastructure, particularly in the construction of tall buildings. There is also a rising focus on intelligent and green buildings in academic areas. This gives us hope for a futuristic vision toward technological innovation and sustainable development. Socio-culturally, Iranian society is strongly inclined to embrace technological adoption and future visioning. Economically, solid Internet security networks and developing intelligent tools by expanding knowledge-based companies will support further research in future research efforts. The second scenario shows latent opportunities along with challenges. Despite facing economic vulnerabilities, Iran has unrealized potential in environmental and historical fields. The existence of laboratories and inactive farmland provides a suitable field for experiments and efforts to protect biodiversity. Meanwhile, abandoned historical and heritage sites provide revitalization and adaptive reuse opportunities. In the socio-cultural aspect, the educational infrastructure in universities provides a platform for cultivating futuristic thinking.

Meanwhile, the socio-cultural dimension's narrative is more prevalent. Because it appears that social trends—becoming increasingly evident—affect how Iranians perceive their future prospects. Although this was an exercise in the context of Iran, it is necessary to provide detailed scenarios for other contexts and gather expert comments through DoA and FD workshops such as IFG. Then, analyze the trends and predictions and make more accurate plans for the future of the cities.

Preparing Iranian cities culturally can be possible by accepting the presented scenarios' society. This method can be made more inclusive by using participatory approaches, particularly those that involve citizens, enhance collaboration between researchers and policymakers, acknowledge cultural values, traditions, and beliefs, forecast their evolution, and incorporate them into scenario planning.

It is essential to mention that to strengthen the process of futures thinking, it is possible to increase awareness through educational programs, public understanding, and cooperation between government organizations, networks, and academic institutions. It is critical to incorporate many viewpoints, listen to everyone in the community, and embrace the most innovative voices and concepts.

References

Ali MM, Al-Kodmany K (2012) Tall buildings and urban habitat of the 21st century: a global perspective. Buildings 2(4):384–423

Asprone D, Manfredi G (2015) Linking disaster resilience and urban sustainability: a global approach for future cities. Disasters 39(s1):s96–s111

Atanasova N, Castellar JA, Pineda-Martos R, Nika CE, Katsou E, Istenič D, Langergraber G et al (2021) Nature-based solutions and circularity in cities. Circ Econ Sustain 1(1):319–332

Bardhan R, Debnath R, Gama J, Vijay U (2020) REST framework: a modelling approach towards cooling energy stress mitigation plans for future cities in warming Global South. Sustain Cities Soc 61:102315

Bellagamba U (2016) From ideal to future cities: science fiction as an extension of utopia. Philos Technol 29:79–96

Bonine ME (1979) The morphogenesis of Iranian cities. Ann Assoc Am Geogr 69(2):208–224

Borkowska K, Osborne M (2018) Locating the fourth helix: rethinking the role of civil society in developing smart learning cities. Int Rev Educ 64(3):355–372

Boudreau J-A (2022) City of repair: practicing the future in Mexico City. Int J Urban Reg Res 46(5):851–864. https://doi.org/10.1111/1468-2427.13102

Boyko CT, Gaterell MR, Barber AR, Brown J, Bryson JR, Butler D, Rogers CD (2012) Benchmarking sustainability in cities: the role of indicators and future scenarios. Glob Environ Change 22(1): 245–254

Cavaleiro de Ferreira A, Fuso-Nerini F (2019) A framework for implementing and tracking circular economy in cities: the case of Porto. Sustainability 11(6):1813

Cheng S, Xie S (2011) Making eco-smart city in the future. In: 2011 International conference on consumer electronics, communications and networks (CECNet), pp 1380–1383, IEEE

Claudio Providas (2023) UNDP Resident Representative in Iran, Iran tackles its cities' carbon emissions, 21 Oct 2021. https://www.undp.org/blog/iran-tackles-its-cities-carbon-emissions

Cruz SO (2015) Alternative futures of global governance: scenarios and perspectives from the Global South. Foresight 17(2):125–142
DESA (2022) World population prospects 2022. Summary of Results, New York
Dezfuli RR, Bazazzadeh H, Taban M, Mahdavinejad M (2023) Optimizing stack ventilation in low and medium-rise residential buildings in hot and semi-humid climate. Case Stud Therm Eng 52:103555
Dezfuli RR, Mehrakizadeh M, Najar BSA, Bazazzadeh H, Mahdavinejad M (2024) Geometric investigation of entrance proportions of houses from the Qajar to the beginning of the early Pahlavi in Dezful City (1789–1979). Front Archit Res 13(1):57–78
Dushkova D, Haase D (2020) Not simply green: nature-based solutions as a concept and practical approach for sustainability studies and planning agendas in cities. Land 9(1):19
Fakhr BV, Mahdavinejad M, Rahbar M, Dabaj B (2023) Design optimization of the skylight for daylighting and energy performance using NSGA-II. J Daylighting 10(1):72–86
Frantzeskaki N (2019) Seven lessons for planning nature-based solutions in cities. Environ Sci Policy 93:101–111
Gall T, Vallet F, Yannou B (2022) How to visualise futures studies concepts: revision of the futures cone. Futures 143:103024
Gearey M, Robertson L, Anderson J, Barros P, Cracknell D (2019) Re-naturing the city for health and wellbeing: green/blue urban spaces as sites of renewal and contestation. Plann Cities Nat Theor Strateg Methods 153–168
Gharaati F, Mahdavinejad M, Nadolny A, Bazazzadeh H (2023) Sustainable assessment of built heritage adaptive reuse practice: Iranian industrial heritage in the light of international charters. Historic Environ Policy Pract 14(4):498–532
Goharian A, Daneshjoo K, Shaeri J, Mahdavinejad M, Yeganeh M (2023) A designerly approach to daylight efficiency of central light-well; combining manual with NSGA-II algorithm optimization. Energy 276:127402
Gravagnuolo A, Angrisano M, Fusco Girard L (2019) Circular economy strategies in eight historic port cities: criteria and indicators towards a circular city assessment framework. Sustainability 11(13):3512
Hara K, Kitakaji Y, Sugino H, Yoshioka R, Takeda H, Hizen Y, Saijo T (2021) Effects of experiencing the role of imaginary future generations in decision-making: a case study of participatory deliberation in a Japanese town. Sustain Sci 16:1001–1016. https://doi.org/10.1007/s11625-021-00918-x
Harrington AR, Stryja M (2021) Eco smart city method of promoting environmental sustainability and sustainable development in an industrialised and digitalized society. In: Internet and new technologies law. Nomos Verlagsgesellschaft mbH & Co. KG, pp 275–294
Heinonen S, Parkkinen M, Karjalainen J, Ruotsalainen J (2017) Energizing peer-to-peer urban futures—challenges for urban governance. Proc Eng 198:267–282
Hiromitsu T, Kitakaji Y, Hara K, Saijo T (2022) What do people say when they become "future people"?—positioning imaginary future generations (IFGs) in general rules for good decision-making. Sustainability 10, 13(12):6631. https://doi.org/10.3390/su13126631
Jamei E, Ahmadi K, Chau HW, Seyedmahmoudian M, Horan B, Stojcevski A (2021) Urban design and walkability: lessons learned from Iranian traditional cities. Sustainability 13(10):5731
Javed AR, Shahzad F, ur Rehman S, Zikria YB, Razzak I, Jalil Z, Xu G (2022) Future smart cities: requirements, emerging technologies, applications, challenges, and future aspects. Cities 129, 103794
Joachim M (2018) Post carbon city-state, rezoned circular economy. http://www.terreform.org/projects_urban-new-york-city-state.html
Kabisch N, Korn H, Stadler J, Bonn A (2017) Nature-based solutions to climate change adaptation in urban areas: Linkages between science, policy and practice. Springer Nature
Khan S, Zaman AU (2018) Future cities: conceptualizing the future based on a critical examination of existing notions of cities. Cities 72:217–225

Kotzen B (2021) Future cities: speculation on the case for vertical biophilic cities. Ecocycles 7(2):59–80
Lafortezza R, Chen J, Van Den Bosch CK, Randrup TB (2018) Nature-based solutions for resilient landscapes and cities. Environ Res 165:431–441
Lakatos ES, Yong G, Szilagyi A, Clinci DS, Georgescu L, Iticescu C, Cioca LI (2021) Conceptualizing core aspects on circular economy in cities. Sustainability 13(14):7549
Lang SB (2007) Predicting or inventing the future? Common grounds in architecture and strategic marketing. Predicting Future 499–506
Langergraber G, Pucher B, Simperler L, Kisser J, Katsou E, Buehler D, Atanasova N et al (2020) Implementing nature-based solutions for creating a resourceful circular city. Blue-Green Systems 2(1):173–185
Langergraber G, Castellar JA, Pucher B, Baganz GF, Milosevic D, Andreucci MB, Atanasova N et al (2021) A framework for addressing circularity challenges in cities with nature-based solutions. Water 13(17):2355
Latifi M, Mahdavinejad MJ, do Vale CP, Iannaccone G (2023) Asymmetric symmetry: an analytical framework of hidden mathematics in Isfahan's valuable Qajar houses; a video-ecology approach
Li J. Yang T (eds) (2015) China's eco-city construction. In: Chapter 1. Ecological crisis, eco-civilization, and eco-cities. Springer
Likitswat F (2019) Future cities: new generation's visions of sustainability concepts and models. Future Cities Environ 5:9–9
Mahdavinejad M, Bazazzadeh H, Mehrvarz F, Berardi U, Nasr T, Pourbagher S, Hoseinzadeh S (2024) The impact of facade geometry on visual comfort and energy consumption in an office building in different climates. Energy Rep 11:1–17
May C, Gandlgruber B, D'Costa A, Botelho AJJ, Balestro M (2022) Brazil and India transitioning to a knowledge economy. In: 34th annual meeting. SASE. https://sase.confex.com/sase/2022/meetingapp.cgi/Session/6024
Miller R, Poli R, Rossel P (2013) The discipline of anticipation: exploring key issues. fumee.org
Miller R (2018) Futures literacy laboratories (FLL) in practice: an overview of key design and implementation issues. Transforming Future 95–109
Moriarty P, Honnery D (2015) Future cities in a warming world. Futures 66:45–53
Najafi Q, Gorji-Mahlabani Y, Goharian A, Mahdavinejad M (2023) A novel design-based optimization method for building by sensitivity analysis. J Solar Energy Res 8(2):1446–1458
Nakagawa Y (2020) Taking a future generation's perspective as a facilitator of insight problem-solving: sustainable water supply management. Sustainability 12(3):1000
Pillkahn U (2008) Using trends and scenarios as tools for strategy development: shaping the future of your enterprise. John Wiley & Sons
Poli R (2017) Introducing anticipation. In: Handbook of anticipation. Springer, pp 1–14
PopulationPyramid, Iran (Islamic Republic of), population in 2050. 10 Feb 2023. https://www.populationpyramid.net/iran-islamic-republic-of/2050/
Riffat S, Powell R, Aydin D (2016) Future cities and environmental sustainability. Future Cities Environ 2(1):1–23
Saadatjoo P, Badamchizadeh P, Mahdavinejad M (2023) Towards the new generation of courtyard buildings as a healthy living concept for the post-pandemic era. Sustain Cities Soc 104726
Saaty TL, De Paola P (2017) Rethinking design and urban planning for the cities of the future. Buildings 7(3):76
Saijo T (2020) Future design: bequeathing sustainable natural environments and sustainable societies to future generations. Sustainability 12(16):6467
Shah H (2023) Beyond smart: How ICT is enabling sustainable cities of the future. Sustainability 15(16):12381
Sharifi A, Murayama A (2013) Changes in the traditional urban form and the social sustainability of contemporary cities: a case study of Iranian cities. Habitat Int 38:126–134

Shichiyakh RA, Klyuchnikov DA, Balashova SP, Novoselov SN, Novosyolova NN (2016) Smart city as the basic construct of the socio-economic development of territories. Int J Econ Financ Issues 6(1):157–162
Shirazi MR, Falahat S (2015) Urban oasis, towards a prototype for sustainable urbanity: the case of Tehran. Int J Soc Syst Sci 7(4):334–351
Shirzadnia Z, Goharian A, Mahdavinejad M (2023) Designerly approach to skylight configuration based on daylight performance: toward a novel optimization process. Energy Buildings 286:112970
Timilsina RR, Nakagawa Y, Kotani K (2020) Exploring the possibility of linking and incorporating future design in backcasting and scenario planning. Sustainability 12(23):9907
UNIDO (United Nations Industrial Development Organization) (2016) UNIDO sustainable cities initiative, hubs of innovation, low carbon, industrialization and climate action, Austria
Uwasu M, Kishita Y, Hara K, Nomaguchi Y (2020) Citizen-participatory scenario design methodology with future design approach: a case study of visioning of a low-carbon society in Suita City, Japan. Sustainability 12(11):4746
Visone S (2020) Eco-Smart governance in the transformation processes of future cities. Masdar city: a model for sustainable cities. Sustain Mediterr Constr Land Cul Res Technol 207–211
Wickenberg B, McCormick K, Olsson JA (2021) Advancing the implementation of nature-based solutions in cities: a review of frameworks. Environ Sci Policy 125:44–53

Part IV
Building Anticipatory Systems in Urban Sectors in Mexico

Chapter 8
Imagining Asbestos-Free Futures—Participatory Mapping as Elements to Anticipate Transitions Based on Hope

Nora A. Morales Zaragoza and Laura E. Quiroz Rosas

Abstract This article reflects on findings about co-creating visions of futures around an asbestos-exposed community in the periphery of Mexico City. It highlights the importance of addressing complex problem approaches from the Transition Design approach in the context of social-ecological systems. We emphasize the need to generate alternative scenarios from the locals that challenge traditional normative urban futures. Our methodological approach was based on interviews and prospective participatory mapping workshops with community inhabitants, who generated visions that reflect their anticipatory assumptions and social dynamics around the problem, allowing them to understand the situation in the present better and to use the future as an asset to promote different levels of interventions considering the spatial dimension. The concept of hope worked as a force reinforced by popular communities' narratives that impulse action toward imagined desirable visions. Finally, we reflect on the need to incorporate spatio-temporal analysis of desires and hopes with the community's inhabitants and other actors to generate comprehensive territorial interventions that contemplate urban-environmental regeneration in broader time frames.

Keywords Socio-ecological systems · Asbestos exposure · Transition design · Participatory cartography · Hope and anticipation of futures

N. A. Morales Zaragoza (✉)
Department of Design Theory and Processes, Universidad Autónoma Metropolitana-Cuajimalpa, Mexico City, Mexico
e-mail: nmorales@cua.uam.mx

L. E. Quiroz Rosas
Department of Social Sciences, Universidad Autónoma Metropolitana-Cuajimalpa, Mexico City, Mexico
e-mail: lquiroz@cua.uam.mx

S. Gonzalez-Arellano and B. Gandlgruber (eds.), *Cities as Anticipatory Systems*, The Urban Book Series, https://doi.org/10.1007/978-3-031-78162-9_8

8.1 Introduction

This article aims to communicate findings related to the development of the capacity to imagine futures on an empirical case of participatory action research involving an asbestos-exposed community in the periphery of Mexico City. From a socio-ecological systems approach, the collective creation of visions from the future within a diversity of actors is crucial to understand complex problems and catalyze change.

Our methodology involved previous interviews among stakeholders and experts from various sectors and a participatory mapping workshop with community inhabitants where they imagined alternative desirable futures related to the problem of asbestos. The first phase of the project, resulted in four preliminary scenarios that confronted views among stakeholders, specially exposing their own anticipatory assumptions and ideas about progress, human development, personal fulfillment, among other aspects. The main confrontation we observed signaled two main perspectives of knowledge, a more deterministic one, that organized human agency toward a more feasible future following the logic of the industrial city, (usually coming from experts and researchers) and a hopeful one, that envisioned a full recovery of health and wellness pointed toward an urban regeneration environment (from the local community).

Our hypothesis considers that the incorporation of the spatio-temporal dimension and the desires and hope from the people participating in the workshop, generated two novel future scenarios reflecting emerging anticipatory assumptions and revealed the social dynamics surrounding the problem, thus allowing a better understanding of the present situation to reconcile the paradigm of anticipating the future and promoting intervention in the present.

The first section of this chapter explains the concept of socio-ecological systems (SES) and Transition Design (TD) approaches, which address the need to generate gradual rather than radical changes in the behavior of society and its relationship with the environment, considering long-term visions toward urban transformations and sustainability. Then we explain the concepts of "transition" and "transformation" which have been central to a variety of current interdisciplinary discourses including some stated in this book, that are concerned on how change manifests and the need to understand dynamic interactions among diverse individuals and communities (see González-Arellano in Chap. 2).

We take both approaches for the action research project of asbestos exposure within an interdisciplinary group of researchers and local participants in order to obtain insight for transformative spaces by creating novel visions of the future and explore how these narratives could create emergent opportunities for change.

The second part reflects upon the current global problem of the asbestos industry, and the impact that it has caused to humanity at an urban scale. We identify some initiatives that the European Commission Act is pursuing to better protect people and the environment from asbestos and ensure an asbestos-free future, which reveals some conscious use of the future and the explicit application of anticipation connecting to future studies. Then we present our case study at San Pedro Barrientos' (SPB) community in Mexico City, and how the integration of prospective participatory mapping techniques worked to generate future scenarios.

In the third part, we analyze the two emerging scenarios from the perspective of hope and its relation to anticipation and the construction of future literacy capacities in the community, explaining how they connect with the notion of urban regeneration (Manzini and Jégou 2003). We also reflect upon anticipatory processes emerging and the type of logic of action and transformative capacity that participants had from the SPB case.

Finally, we conclude that more exploratory approaches involving visual socio-spatial methods are needed to co-create future visions with a diverse group of actors in the context of socio-ecological systems, which could contribute to creating awareness and scale-up actions toward complex global environmental problems.

8.2 Socio-ecological Systems and Transition

The global crises and problems caused by the global environmental change that characterizes the Anthropocene (Steffen et al. 2011) have stimulated interest in science, policy, and society for change from a systems approach (Folke 2006).

This new geological era presents essential environmental and social challenges, such as the depletion of natural resources and increasing inequality and social vulnerability, which require research and innovative approaches to understand the dynamic interactions between ecosystems and human societies (Becker 2012). Socio-ecological systems (SES) approach refers to complex adaptive systems that require a holistic understanding and an interdisciplinary approach to study and address their challenges (Preiser et al. 2018).

Despite some academic efforts to integrate complex adaptive systems into SES research, such as cities, there has been limited reflection to identify relevant methods and practical approaches to study SES, specially about intentionally guiding transition toward more sustainable infrastructures and economies. Given this lack of clarity, a common thread is needed to address the threats we face in an equitable, fair, and sustainable manner that recognizes the need to *transition* and *transform ourselves* in order to create fundamentally different social-ecological interactions at all levels that reduce inequality and social vulnerability (Gunderson and Holling 2012; Walker et al. 2004; Folke et al. 2010; O'Brien 2022).

The terms "*transition*" and "*transformation*" have emerged in political and scientific discourses, which have been used to refer to radical, large-scale, nonlinear, and structural changes in complex adaptive systems to generate a sustainable society (Feola 2015; Patterson et al. 2017). Both concepts provide nuanced perspectives on how to describe, interpret, and support social change. Their differences are due, on the one hand, to their etymological origins and, on the other hand, to the use they have been given by different research communities interested in *transition* or *transformation* (Hölscher et al. 2018).

Transition has been used mainly to analyze changes in specific social subsystems (energy, mobility, cities), maintaining a systemic view between technical, environmental, and institutional interactions (Loorbach et al. 2017). *Transformation,* on the other hand, is more commonly applied to refer to large-scale changes, with short and long-term actions that can be global, national, or local and involve the interaction of the components of the socio-ecological system, integrating concepts such as resilience and planetary boundaries (Folke et al. 2010). In the following (Table 8.1), some premises of coincidences and differences between both concepts are presented, emphasizing the need to apply integral approaches for their effective implementation at different scales in the system, from normative or regulatory futures, usually taken by formal agreements between governments, plans, or laws that require compliance with certain behavioral patterns as in the case of the Global Goals for Sustainable Development (González-Arellano 2022). As well as local and even individual initiatives within existing conditions that opposed imposed processes and perspectives.

In addition to the operational differences outlined above, this paper used methodologies drawn from both concepts to expose the case study: transformational spheres (O'Brien and Sygna 2013), social cartography (Diez Tetamanti and Chanampa 2016) and Transition Design (Irwin et al. 2015).

Table 8.1 Comparison between applications of terms of "*transition*" and "*transformation*" concepts adapted from Hölscher et al. (2018)

Scale of change	Transition	Transformation
System approach	Complex adaptive systems	
	Social, institutional, and technological change in social subsystems	Processes of large-scale social change (global, regional, local)
Dynamics and processes	Identifies patterns and mechanisms of complex systems, including dependency, emergence, and thresholds	
	Describes how nonlinear change occurs, focusing on the dialectic between limits and opportunities	Attempts to explain emerging patterns of change and the impact of outcomes
Regulations	Change is questionable and may be desirable or undesirable	
	Results focus on the change from unsustainable to sustainable system status	The results focus on creating safe and fair operating spaces, avoiding undesirable changes in the system
Agency and Governance	Multi-stakeholder process that enables innovation, learning, collaboration, and knowledge integration	
	Development of disruptive interventions to support sustainability transitions	Respond to the implications of change (e.g., risks, vulnerabilities); individual motivations and values that support transformations

8.3 Transition Design and Future Studies

The TD framework developed by Irwin et al. (2015, 2020) recognizes that we are living in "transitional times" and takes as its central premise the need to catalyze social transitions toward more sustainable and just futures, with the understanding that design has a key role to play in facilitating the performance of these transitions. It is based on a heuristic structure around four interrelated areas with some unique features. This design approach is linked to a growing awareness of a systemic problem and analysis with multi-stakeholder involvement and a clear orientation toward longer time horizons explicitly informed by visions of sustainable futures (Escobar 2016).

After the understanding of the problem has been explored with different participants, the phase of co-creation common preferred futures follows by using *backcasting,* a methodology used in the future studies, developed by Karl Dreborg (1996), that focuses on imagining and planning for desirable and sustainable futures, and then working backward to identify the actions needed to achieve those futures. Unlike forecasting, which focuses on predicting the future based on past and present trends, backcasting begins with a vision of a preferred future and then analyzes how to get to that future from the present.

The collective generation of those visions in long-term futures forces the stakeholders to identify the more problematic structures in shaping their pathway to navigate transition. Some articles explain the adaptation of these techniques in various projects such as agroecological transition (Morales Zaragoza et al. 2021), biocultural change of cancer (Valadez-Blanco et al. 2020), water resilience (Morales and Barraza 2020), and sustainable tourism in Latin American contexts.

In the Asbestos case, while four scenarios (described ahead) were already developed following the DT approach, we emphasize the use of the "Prospective Social Mapping Workshop," conducted only with local community participants, which generated two novel future scenarios, and their emerging anticipatory assumptions that unveiled social dynamics of the local community surrounding the problem, this allowed to better understand the problem by integrating a more holistic view of the factors involved and promote different scales of interventions for the project.

The next sections we stress the use of this participatory mapping role in the construction of future visions using local participants' own imagination and suggest something that could be added to the DT model. We would add a complementary analysis from a socio-ecological system perspective by incorporating the notion of hope and well-being as anticipatory forces in relation to a vision of regenerative and the eco-sanitary urban future. We draw this notion, from future studies literature and particularly the work of anthropologist Rebecca Bryant, who incorporates an affective and emotional dimension to the concept of anticipation and hope, drawing a distinction to think about it relation to the past; she suggests that a thick connection gives us insight into what she calls *teleologies* of action.

8.4 The Global Problem of the Asbestos Industry and Its Impact on Health and the Environment

Asbestos has always been a highly dangerous, cancer-causing substance, the World Health Organization considers it a hazardous and deadly material affecting thousands of people worldwide (World Health Organization 2015). Despite being confirmed as a highly dangerous carcinogenic substance and banned among several countries, asbestos fibers are still present in the urban environment and are responsible for many preventable deaths around the world (European Commission 2022) as they have been used for decades, mainly in the construction industry. Ideas of globalization and development have had important effects in supporting commercial strategies of the multinational asbestos industries, introducing them to emerging economies. Even governments have been the leading promoters of this material due to the demand for economical and heat-resistant building materials for the development of cities despite these well-known dangers. This practice has inevitably been used worldwide in many public and residential buildings (Wallis et al. 2020).

There is epidemiological evidence confirming that all types of asbestos and its fibers are implicated in the causes and development of various diseases (such as mesothelioma, lung cancer, asbestosis, pleural thickening, and pulmonary plaque) and premature deaths. Another surprising aspect is that 30 years after the asbestos boom in the 1970s, the true extent of this hidden danger is exposed (Wallis et al. 2020).

Although many countries have achieved a total ban on the use and manufacture of asbestos, unfortunately, the history of regulation shows that efforts are insufficient, the processes are slow, and sometimes corrupt to mask the risk control of the material itself with a false promise (Hadley and Rennell 2013, p. 36). In that sense, the commercialization and use of this material have prevailed over human, environmental, and labor rights. Despite advances in the regulation and prohibition of the use and manufacture of asbestos products in the countries of the Global North (Canada, the United States, the European Union, and Japan), the countries of the Global South in Latin America, Asia, and Africa, including Mexico, continue to lag in this area.

Taking the case of the European Union, even though all forms of asbestos have been banned since 2005, it is still present in older buildings and represents a threat to health when asbestos-containing materials are altered, and fibers are released and inhaled, for example, during renovations. Therefore, the European Commission (EC) for the Prevention and Control of Cancer[1] (European Commission 2022) has generated a program in 2005 that seeks to modify labor policies to protect the environment and people from asbestos exposure and it is considering how to ensure an asbestos-free future. It addresses three advocacy priorities in its objectives:

- The search for improved diagnosis and treatment of diseases caused by asbestos.
- Through the detection and safe removal of installed asbestos.

[1] The commission comprises member states of the European Union EU and workers and employers from the asbestos' industry.

- To treat waste containing the material.

8.5 Environmental Issues and Future Studies

These types of environmental issues initiatives concern a particular interest in futures studies, which, according to Miller (2018), leads us to explore the concept of "futures literacy" (FL) as a tool for understand collective anticipatory processes and systems that project actions within the imminent and rapid urbanization of the world, which primarily can be used to innovate in the form of urban organization and foster decision-making within a wide diversity of problems and risks.

Understanding the nature of the future of environmental, human health, and toxicity is related to the development of the capacity to imagine actions to catalyze change today. The example of this European plan demonstrates how simply thinking about what might happen with the asbestos installed in buildings and ask ourselves how it will directly affect the way society will live in the future can lead to developing anticipatory skills and improving the design of collaborative processes (Miller 2018, p. 8) this complex system, can support the capacity building of people affected by a problem to "break free" (Nussbaum 2013), and transform their reality.

The main rationale of "Futures Literacy Framework" (FLF) emerges as a foundational theory for exploring how people perform anticipatory awareness activities and provides an analytical ground for mapping these activities in search of solutions to global problems that no longer work with current systems and leads us to explore new ways of approaching what we want to change about the dominant regime. The three aspects of the free-asbestos program proposal show a clear line of planning and preparation as an instrument to get to a future goal, but how about new kinds of futures for different purposes. Is there a way to look for more alternatives?

8.6 Hope and Anticipation in the Construction of Futures

In the book "*The Anthropology of the Future*," Rebecca Bryant and Daniel Knight propose an innovative approach based on six orientations or domains of the future (Anticipation, Expectation, Speculation, Potential, Hope, and Destiny) which are interrelated and show ways the future may orient the present, and shape the way people experience and think about the future and how these can affect the present (Bryant and Knight 2019). Each orientation can be analyzed from social discourses and practices in different societies.

We picked anticipation and hope from Bryant's anthropological perspective, because she examines how humans imagine, construct, and relate to the future, both individually and collectively, suggesting that anticipation and hope are not only mental states, but also social practices rooted in specific cultural contexts.

Both orientations are fundamental components of the human condition, shaping our actions and perceptions of the world around us. Bryant theory explores how these notions influence various aspects of life, such as urban planning, survival strategies, politics, and religion. She provides a valuable theoretical basis for understanding how we imagine and construct the future.

For Bryant, hope is a dynamic force that drives action and imagination toward the future, shaped by specific social and cultural contexts and with important ethical and political implications (Bryant and Knight 2019). This orientation holds on to the belief that the future will be better, as it is shaped by people's present experiences and the possibility of returning to specific standards or ways of life that motivate them to act in the present.

Anticipation is seen as a form of future orientation that drives people's action and imagination. This orientation to the future is seen as a fundamental human capacity that involves preparing for the future, considering different possibilities and facing the uncertainty and anxiety that accompany the expectation of the unknown and involves these key features:

1. Preparation and planning: Anticipation involves the ability to anticipate future events and prepare for them. This may include planning concrete actions to achieve goals or avoid risks.
2. Expectations and possibilities: Anticipation involves having expectations about what might happen in the future and considering different possibilities and scenarios. This ability to imagine alternative futures is fundamental to anticipation.
3. Uncertainty and anxiety: Anticipation can be accompanied by uncertainty and anxiety about what the future holds. Bryant recognizes that anticipation can generate both positive and negative emotions, depending on individual circumstances and expectations.
4. Cultural and social context: Like hope, anticipation is shaped by the social and cultural context in which it is found. The ways in which people anticipate the future may be influenced by cultural norms, social expectations, and past experiences.

On the other hand, a characteristic of hope is the collective sense of experiencing a critical juncture in which something is happening, usually accompanied by a time of previous austerity or promises that have not been fulfilled. Hope begins as something lost with some restorative or prideful power taken from the general public or nation (Bryant and Knight 2019). In that sense, it can be compared to our case since the perception of having lost one's living environment due to the intervention of the industry specifically related to asbestos contamination, but extrapolated to the current presence of other industries such as cement or the vast warehouses and storage buildings from the speeches of the workshop participants or informants interviewed from the community, denotes a sense of helplessness that plunges them into a state of discouragement as they have no possibility of recovery but at the same time the promise that they can recover the lost standards as citizens or the lost desire to reconfigure themselves as a community.

8.6.1 Hope as a "Latent Moment" of the Future

Taking it further, Ernst Bloch's relates hope to a kind of **anticipatory consciousness** in his book "The Hope Principle" (2007)[2] where he explores utopian futures and describes a moment "yet to be realized," which could also be explained as a kind of "exaggerated anticipation perpetrated by a kind of desirable imaginary reinforced by popular stories, fairy tales, fiction, theater, and other mass media." Bloch, based on Schatzki's notion of "practice-arrangement bundles" that create their time–space structures, tells us that these imaginary (produced, consumed, and shared) functions as a kind of "affective temporal spaces" or *teleoaffects* that influence our future actions. The "*teleoaffections*" refer to the set of accepted images that combine with the technologies of the imagination to produce a kind of "latent moment" that attracts us to "something" indeterminate (Bloch 1995, p. 18).

For Bloch, hope is a future orientation that functions as a bridge that connects the gap between potential and actuality, and its value lies in contemplating the future by allowing us to think about possibilities; it asks us to move away from the focus on the present state of social relations and focus on what they may become.

Both Bloch, Bryant and Knight (2019) relate the domains of hope and potential by stating, "The immanence of not-yet-realized potential becomes a resource of hope," detailing that when a potential has not been realized or is frustrated, there is the possibility of the emergence of hope, providing the "momentum" of mobilization and momentum toward actualizing the future.

One quality of hope worth mentioning is the notion of "perpetual motion," seen as a potential tendency that propels toward something; hope arises in the gap between the potential and the actual and can guide us toward the future even if it is perceived as an insignificant force, it can be built as an unstoppable force. What may begin as a gathering of collective desires can be transformed into seeds of hope that seek to materialize the form of what does not yet exist.

In this sense, we can affirm that the scenarios created by the local participants reflect a sense of collective hope as a community, starting from the moments of daily life emerges an uncertain view of "of what has not yet been conquered, nor has it been completely won."

It is important to reiterate that the reinforcement of the historical potential of the past is a critical aspect of improving the mobilization capacity of the collective. However, the low credibility of the participatory methodologies, the constant time pressure of the project, as well as the lack of openness of the research actors to include aspects that involve the imagination and creativity of the people as drivers of citizen action, and the presence of an overtly political orientation of the project, led to the conclusion of the other social mapping workshops with the community.

If hope is nourished by indeterminacy (or ambiguity, as the case may be), it is also rooted in potentiality and the possibility that it concludes with the impossibility of realization. Therefore, some anthropologists, such as Bloch (2007), suggest that it is

[2] The philosopher's original work was written in 1938 during his exile in the United States of America and later reworked after his return to Germany in 1947.

essential to study it alongside disappointment, a way in which potential is partially actualized.

According to Bloch, the first level of disappointment is that the "here and now" is close to the present. The raw experience that brings us from the drifting dream to another state: that of immediate proximity, where the recognition of reality is more complex and obscure than the "desirable image."

Disappointment is the recognition that the actualization never considers the potential or the paradox of what can happen but the realization "that the dream seemed firmer or, in any case, brighter than its realization" (Bloch 2007).

What builds hope is precisely the idea that ways of life do not necessarily coincide with the existing reality but the surplus potential of future orientation, which provides the motor or movement toward that indeterminate "something."

In his book *The Methods of Hope* (2004), Miyazaki argues that hope is an indeterminate orientation to the future and a method of living. People displaced from their land in 1880 in Fiji continue to seek compensation from the government despite continuous rebuttals and failures. Drawing on ethnographic theory and taking up Bloch's concept of indeterminacy, he criticizes the Greek-derived assumption that humanity is captured by the notion of history as a theological course of truth about the essence of life.

The future does not unfold according to the planned or imminent teleology.[3] The future will always surprise and stimulate us. Indeterminacy is at the heart of the theory of hope as a method.

8.7 The Case of Asbestos Exposure in San Pedro Barrientos

The participatory action project that frames this research work, entitled: "Transdisciplinary approach to the epidemic of pleural malignant mesothelioma due to asbestos exposure in the community of San Pedro Barrientos (SPB)" is funded by the Mexican National Council of Science and Technology and aims to influence polluting processes policies of toxic damage as well as address the socio-environmental impacts associated with sources of natural and anthropogenic origin.

San Pedro Barrientos is located in the north part of the metropolitan area, the municipality of Tlalnepantla de Baz, one of the most industrialized areas in Mexico City. In the early 1940s, one of the leading Asbestos Mexican factories settled in the neighborhood, with the promise of economic development and well-being of their inhabitants (mainly family workers), after 58 years of the final factory departure in 1998. The place was left as a polluted environment and a poor and socially fragmented population of less than 6000 inhabitants survivors of exposed workers families and

[3] In the context of Rebecca Bryant's work, "teleology" refers to the study of ultimate purposes or goals in nature, society, and history; it involves the understanding of how people imagine and strive to achieve certain goals or end states in the future.

an increase in pulmonary malignant mesothelioma cases associated with asbestos exposure (Gopar-Nieto et al. 2016).

Although the project was approved in February 2022, the researchers began to organize and develop a previous diagnosis with experts, community inhabitants, and other stakeholders a year before, in remote mode, due to the confinement of the COVID-19 pandemic that started in Mexico in March 2020. The initial stages of research (following DT methodology), involved academic actors, institutions, activists, and community leaders understanding the complexity of the asbestos problem in context of ecological health from a multilevel perspective. The participants explored historical events that impacted the community and region during the years, such as statistics about the pleural malignant mesothelioma cases of the exposed population, continued with the identification of historical landmarks and community practices and activities that reflect their regular lifestyles until today. After some research based on interviews with experts and secondary reviews that put into context the dimension of this toxic impact, some co-creation workshop sessions were held, in which the interdisciplinary group came up with four preferred desirable future scenarios, which we describe next:

Scenario 1 "Recovery of green spaces and isolation of the endangered zone." This scenario situates San Pedro Barrientos community in a medium-term future, a space where green areas are recovered with trees and plants among residential areas, the areas exposed to contaminated dust have been identified and isolated by a thick wall from the people. The inhabitants have improved risk management measures due to well-informed programs invested in screening and early diagnoses held by the government and citizens. Local employers and businesses are organized in a commission responsible for risk management and protecting inhabitants against exposure.

Scenario 2 "(S.O.S) A call for help to the urban authorities." This scenario captures a tone of urgency within the problem of contamination and asbestos exposure, covering a short-term vision. It is an urgent call for "rescue actions" from the authorities for the victims of asbestos exposure, similar to situations of natural disasters (flood, fire, or earthquake), where people affected are quickly "extracted" from the site and relocated to safeguard temporal areas. This scenario also contemplates support for victims of asbestos-related diseases.

Scenario 3 "Total ban of asbestos in Mexico." This scenario, also situated in a short to medium temporal span, shows a Court government legislative congress scene where Mexico has banned all uses and products of asbestos. This regulatory-oriented future does not necessarily contemplate detailed implementation law issues, such as the cessation of importation and use or the agreements of "controlled use" or "asbestos substitute materials" still being used in the construction industry. It does not consider funding available to support health prevention or removal of asbestos materials in buildings and other urban infrastructure.

Scenario 4. For a future free of environmental pollution. This scenario envisions a long-term future where the inhabitants of San Pedro can walk freely through the streets and engage in sports and recreational activities without worrying about air pollution or the environment. The former factory and industry site is now a center for asbestos research.

These scenarios caused some concerns and discomfort among the scientific and academic researchers of the project, because they were considered improbable and not feasible for achieving solution initiatives from the locals, some expressed their concerns and irresponsibility to create false hopes or expectations for the inhabitants. This reaction was in part natural due to the lack of understanding of the backcasting methodology and our limited experience to deal with experimentation and imagination processes to generate new innovations and visions of the future.

According to Riel Miller (2018), the above situation is a common aspect that arises when people "use futures," and confronts two types of anticipatory systems; one that he calls **anticipation for the future** (AfF), which refers to a type of plan or desired future to which people bet based on the probability of happening (which refers to the researchers assumption based on their current information and knowledge perceived as improbable to realize). The other type is **anticipation for emergence (AfE)** which contemplates emergence as a kind of "non-future" or alternative option to the dominant perspective, here the important thing is not having a goal to reach or achieve plan or preparation, but to be able to imagine something beyond-now, to create a disposable construct, which need not be limited by probability or desirability. As Miller puts it, (AfE) is about a strange future that allows us to feel and make sense of present aspects with some novelty that will always be overshadowed by anticipation for futures.

At this point of the project, we wanted to pursue other ways to incorporate the local community in the generation of desirable scenarios so we conducted a "Prospective Social Mapping Workshop" following Diez Tetamanti and Chanampa work on social mapping (2016).

8.7.1 "Social Cartography" Workshop with the Inhabitants

Participate social cartography is a collection of methods, attitudes, and beliefs that allow the people of a community to express and analyze the conditions of the reality in which they live through the collective production of maps, with the objective of planning, taking actions, monitoring, and evaluating their own results in a territory (Chambers 2006). The use of this tool allows, among other things, to involve and empower the community in decision-making, in addition to raising awareness about a conflict. This mapping method also adheres to the proposal Deleuze (2017) who argues that subject and object are not divorced, and the communion of hands amalgamates in the process of shared and mobile experience. This process of experience places both the researcher and the object of research on a common plane where the experience is the process that unifies them in the dialogue and exchange of information, by which both the research subject and the researched subject share the experience.

Fig. 8.1 Three participants in the prospective social mapping workshop in San Pedro Barrientos

The prospective social mapping workshop[4] was held on September 22, 2022, at San Pedro's Public High-school (Fig. 8.1) to complement the future visions generated in the research project and produce a collective map that communicates the experiences, activities, and perceptions around the daily spaces of the inhabitants of the community, encouraging reflection on the legacy of the past, current concerns, and needs for future generations.

The workshop was attended by seven participants from affected families in the area. We began by giving a guided presentation of the activity's objectives in which we introduced the research initiative, the local demand for the project, and the need to coordinate other intervention events in the territory.

Three key questions were posed: How do people live in SPB today? How did they live in the past? What are the main concerns of the community members? What challenges do they face, and what desirable futures do they aspire to?

[4] Diez Tetamanti and Chanampa conceive social cartography as a productive process that involves a succession of events and phases around the conjunction of collective representations and transformations from the production of new spatial senses (Diez Tetamanti and Chanampa 2016).

8.7.2 Workshop Structure

Stage 1. Base map Construction. Location of prominent spaces in the community (public spaces, streets, schools, and factory areas).
Stage 2. Spaces of the past to be intervened (existing objects, current conflicts, social practices, and social relations).
Stage 3. The Envision future. The imagination of a desired future, incorporating new objects, conflicts, and practices, and looking for relations between objects and society.

8.7.3 Workshop Key Findings

The workshop fostered an atmosphere of dialogue and information exchange between participants and researchers, resulting in three maps (Fig. 8.2) that gave feedback for two novel future scenarios and five key findings for the project:

K1. Political segmentation. Inhabitants used to join political parties because they hoped to influence their community and create a reputation of leadership; however, after corruption and deceit, they became divided and tried to stay away from politics; leaders of the colony looked out for their benefit.
K2. Decrease in recreational and convivial places. Recreational areas in the neighborhood, such as soccer fields and sports fields (which were promoted by the factory in the past), have been privatized, generating an economic barrier to the organization of collective sports activities such as soccer. Today, it is easier and cheaper to organize activities outside the colony. We highlight the reflection of one participant: "Industry and development have caused a series of damages to our community; they have taken away recreational areas, where we used to walk on Sundays or play sports, the places where we used to sit and live together, are now poisoned places."
K3. Localized contamination. Most of the workshop participants identify the contaminated areas near the location of the old asbestos factory, and others, such as those near the roads, speak of the disappearance of entire families due to asbestos-related diseases. They also mention a change from the establishment of industries such as the cement industry (which boosts the labor economy but continues to be polluting) to storage warehouses that bring greater insecurity because they are perceived as abandoned areas or do not allow the flow or connection of the community. A key point of community concern is the workers' union building (constructed entirely of asbestos), which some workers hope to sell. The soil contamination study with the presence of the contaminant corroborated all of these perceptions.
K4. Change from pedestrian to vehicular mode, affecting accessibility and safety. The increase of fast roads is a common and naturalized phenomenon of urban growth, causing isolation and connectivity between neighborhoods and

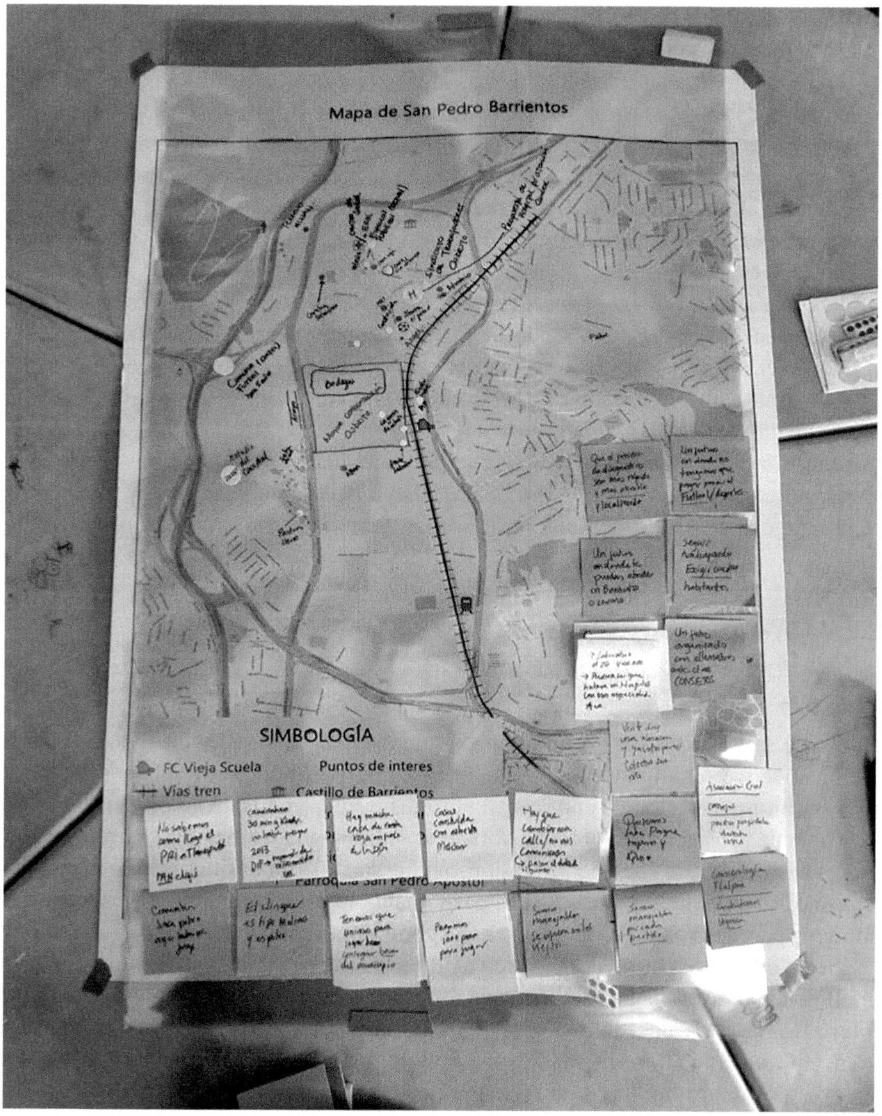

Fig. 8.2 One of the maps resulted from the activities of SPB's Workshop. Depicted as one of the main proposals for the future of SPR, we can see The General Cancer Hospital in the same place as the Workers Union building

populations; the large roads become unbreakable borders, practically preventing walkability and increasing the recurrence of road accidents. Workshop participants were concerned about the increase in the number of people being run over. Participants recurrently talked about their concerns about walking safely in streets either to do an errand to a grocery store or the market or to walk their children back from school (some participants noted that those trials have increased in distance since there were fewer schools for SPB children). They point out that neighbors tend to remove trees and build fences, diminishing accessibility and coexistence. Inhabitants tend to normalize the confinement state (reinforced by the pandemic) but also react to the lack of security and the increase of drug dealing in the area; this phenomenon also affects the low walkability in the streets and the problems of external migration and the increase of vehicular load.

K5. Preferential health access for the population. It is based on a story about the experience of a young participant in the workshop who told how he was diagnosed with prostate cancer and followed his treatment, going every 15 days to the cancer hospitals in the south of the city. Thanks to this, he was in remission. However, he had just received the news of lung cancer. He had decided to give up social security to accelerate his treatment and to look for alternative economic means impacting the family economy since this "new affectation" forced him to return to the beginning of a very complex and delayed process that represented the difference between life and death. They also commented on the problem of shortage and cost of medicines that forces them to access the black market.

8.7.4 Relationship Between Workshop Key Findings and Visions of the Future

Table 8.2 synthesizes the four scenarios previously generated with the findings of the prospective social mapping workshop, which reveal different values and shared beliefs among participants and researchers (Chevalier and Buckles 2019).

The scenarios generated by the workshop participants were taught from everyday life experiences and the incorporation of the spatial dimension, which brought a tangible component that encouraged participants to project an image for the future, allowing people to express themselves and communicate their ideas, fears, and hopes.

We could say that this workshop led facilitators and participants shared a collective intelligence process that uses the future differently. When a lack of knowledge of the future is assumed, a certain equity of ignorance forces reliance on the imagination to invent images of tomorrow. This methodology helped to gather evidence on how "people use the future" and, at the same time, citing Latour (1999), allowed us to learn from the inhabitants without imposing an a priori definition of the world and its capabilities.

Table 8.2 Scenarios linked to key findings from the social cartography workshop

Scenario	Findings to address/retake or reinforce	
S1. Recovery of green spaces and recreation	K2 K4	*Decrease in recreational venues.* This future is linked explicitly to the recovery of sports and recreational activities that are valuable and meaningful to the community (such as soccer) that strengthen social cohesion and reinforce the sense of identity *Change from pedestrian to vehicular mode, accessibility, and safety issues*
S2. (S.O.S) A call for help to urban authorities		
S3. Total ban on asbestos in Mexico		
S4. For a future free of environmental pollution	K3	*Localized contamination.* Where **remediation** and **detoxification of sites contaminated** by the mineral have been achieved, this aspect has caused greater concern and rejection by the scientific actors of the project since they understand the low feasibility and the long-term frame involved in such an initiative. Few case studies from industrialized and island countries demonstrate the potential of bioremediation or regeneration of massive amounts of soil contaminated by asbestos. Their results are not encouraging as they are influenced by cost and land availability (Wallis et al. 2020) Despite having few feasibility options in terms of initiatives, it reinforces Meadows' (2008) argument as it implies a rethinking of urban collective infrastructure that is still on the distant horizon

(continued)

Table 8.2 (continued)

Scenario	Findings to address/retake or reinforce	
NEW SCENARIO S5. Inhabitants with expertise in regeneration and removal of installed asbestos	K1 K3	*Political segmentation.* A workshop key finding could be defined as "**Community vs. political support,**" participants imagined a future that responds to the need to explore **alternative forms of organization** that do not depend on political affiliation to a particular party and the need to have alternative leaderships of socially committed citizens, who have recognition by the community and focus on everyday demands of welfare and health instead of individual benefit When exploring the exercise of imagination toward more distant temporal futures, common representations of the future emerge framed with the question: What would have to happen to the inhabitants of San Pedro Barrientos to reach this scenario? An interesting concept is related to the education and training of the inhabitants themselves regarding asbestos, first by establishing an asbestos museum and later by becoming **experts in remediation and care**; this new alternative opens up occupational and economic opportunities for the community (we can also extend this to the concept of citizen science)
NEW SCENARIO S6. The National Medical Cancer Center at SPB	K5	This scenario arises from the question posed in the mapping workshop from finding **H5**. *Preferential health access for the population.* In a few decades, the population of San Pedro Barrientos has changed from being an industrial corridor to a health complex specializing in cancerology; the first hospital was built in the location of the Asbestos Union building itself that the government bought from the workers and took care of removing the asbestos installed in a safe way for the population

8.8 Conclusions

We have addressed the SES and Transition Design (TD) approaches on describing, interpreting, and supporting desirable nonlinear social changes and complexity to facilitate a collaborative exploration of future scenarios in regard to a free-asbestos future. While doing so, we were able to identify diverse ways of practicing anticipation at the individual and collective levels.

To answer the question in regard to agency improvement and collective anticipation capacities in regard to the urban group of actors in the asbestos case, we were able to identify two kinds of anticipation: the first stage contemplates more the idea of anticipation as an activity that reinforces planning and programming toward the desired future. In fact, the discomfort reactions of the group of actors evidenced internalized anticipatory systems that are more normative and anchored to a way of creating knowledge. This way is difficult to modify at a conscious level, since the agency is framed in a series of decision-making and responsibilities that go according to a vision of the future that is preferable as a "goal to reach."

In that sense the first four scenarios generated for the research project helped us better understand the various social dynamics in the community around the problem and how to reconcile the present situation within different paradigms of anticipating the future. As Reed et al. stays, (2018): "Future scenarios generation processes go hand in hand with modes of engagement with stakeholders and span a continuum from information to consultation to co-production of transformative visions."

However, we can highlight that the participatory mapping workshop and the identification of hope as an orientation toward the future, gave an account of a different way of understanding anticipation. This second type of anticipation that Miller describes as: anticipation for emergence refers to making sense of aspects of the present, and the emergence of novelty. This type of anticipation relates to the ability to engage in improvisation and the acquisition of knowledge necessary to make sense of complexity including one that it's crucial specificity-unique and ephemeral (Miller 2018).

Based on the former, we could say that the prospective participatory mapping workshop, was a crucial step for the participants from SPB in regard of their literacy future capabilities, because even coming from already socially fragmented community they were able to embrace complexity rather than just lamenting it as some cursed and inescapable situation.

This tool in particular allowed the participants to project two novel desirable future scenarios for the research project. Which situates the potential of the spatial–temporal dimension of the creation of new images for the future and fostered some kind of anticipatory awareness of a potential possible future by allowing the participants to project their expectations.

Although the whole methodological experience was not able to generate concrete actions from the local actors in regard to a free-asbestos future, (that was part of the responsibilities of the research group). Following Miller's description of the Future Literacy Laboratories (FLL) model, which is seen as a continuing action learning process where people engage in a knowledge creation process designed to imagine emergent futures. We were able to identify three emergent capacities:

(1) Participants became explicitly conscious of how the future plays a central role in what they perceive and pay attention to, in the present, even if they do not immediately benefit from it.
(2) The capacity of imagining different futures lets participants be aware of their own capacity to invent and identify underlying assumptions. They became better at rooting their own history and specific socio-economic-cultural context.
(3) Participants started to reassess their perceptions of the present, depictions of the past and aspirations for the future.

Another assertive aspect that we would like to highlight, is the use of Bryant orientation of hope as complementary to anticipation in the asbestos case. “Hope emerges as a value that promises a new prosperity on the horizon”; this new horizon becomes feasible when the possibility of regeneration of the inhabited space becomes plausible from the direct message that reflects people’s expectations.

The two new scenarios created by the community summarize a desirable or dreamed future and demonstrate the potential of a future blocked by industry and other actors like authorities discourses among years that reinforced a promise of economic development that has yet to be realized today.

Based on hope, the anticipatory assumptions reveal a potent past, present, and future narrative within local communities that can feed interventions at different system scales, functioning as acupuncture. These imaginaries are the first steps toward collective remediation actions by the community’s inhabitants and could potentially transform into an urban regenerative public policy by influencing decision-makers and authorities.

We also agreed with Bryant when she argues that “the future often represents an unknown against which persons struggle to maintain stability clung to particular stories,” that is what happened when the research and academic group felt confronted by the desired scenarios generated by the community.

We could conclude that anticipatory capacity is a flexible process that involves various actors at different scales. However, in Latin American cities, the asbestos situation is not an anticipatory remediation system, as it is characterized as a reactive, prohibitive, low agency, and zero regenerative system. Somehow, the generation of these new scenarios (S5) regenerative future and economic alternative and (S6) a cancerology center in SPB opened new possibilities to explore a regenerative urban aspect from top-to-bottom. The above reinforces to Bloch’s concept of “dormant potentials,” which in a way were triggered by the workshops, and benefit both the individual and the collective in the project.

Despite some challenges and barriers related to improving a more reliable structure that overcome restrictions and limitations of working with communities and integrating emotional aspects in foresight practice. We propose the need to explore tools that allow a diversity of actors within the three transformational spheres to imagine futures.

We believe our work adds up to the analysis of how societies and individuals conceive and work toward the future, and how these visions of the future could influence their actions and decisions in the present. We stress the importance of developing tools that facilitate collective imagination that could lead society to change or generate insights about different values and aspirations that come into play in societal systems at different scales.

Finally, we posit that the importance of developing anticipation capacity is perhaps the ultimate promise of future studies practice, which not only treasures and preserves expert findings and tools but also provides user-friendly generative tools that facilitate the expression of a larger population. Stuart Candy’s suggestion of using experiential futures and, specifically, playful activities may help enhance this promise of democratizing tools (Candy 2018).

References

Becker E (2012) Social-ecological systems as epistemic objects. pp 37–59
Bloch E (1995) The principle of hope, vol 1 (N. Plaice, Ed.; 1. MIT paperback ed). MIT Press
Bloch E (2007) The hope principle/Bloch, Ernst. 1: The hope principle [1]: Written in the United States between 1938 and 1947. Revised in 1953 and 1959. Edited by Francisco Serra. Translation by Felipe González Vicén. Trotta
Bryant R, Knight DM (2019) The anthropology of the future. https://doi.org/10.1017/9781108378277
Candy S (2018) Gaming futures literacy. The thing from the future. In: Miller R (ed) Transforming the future. Anticipation in the 21st century, 1st edn. Routledge, pp 233–246. Gaming Futures Literacy|7|The Thing from the Future|Stuart Candy
Chambers R (2006) Participatory mapping and geographic information systems: whose map? Who is empowered and who's disempowered? Who gains and who loses? E J Info Syst Dev Countries 25(1):1–11. https://doi.org/10.1002/j.1681-4835.2006.tb00163.x
Chevalier JM, Buckles DJ (2019) Participatory action research: theory and methods for engaged inquiry, 2nd ed. Routledge. https://doi.org/10.4324/9781351033268
Deleuze G (2017) The anti-oedipus: capitalism and schizophrenia (1st ed, 1985, thirteenth printing). Paidós
Diez Tetamanti JM, Chanampa ME (2016) Perspectivas de la Cartografía Social, experiencias entre extensión, investigación e intervención social. Revista +E Version Digital 6:84–94
Dreborg KH (1996) Essence of backcasting. Futures 28(9): 813–828. https://doi.org/10.1016/S0016-3287(96)00044-4
Escobar A (2016) Autonomía y diseño: La realización de lo comunal (1st ed in Spanish). Editorial Universidad del Cauca
European Commission (2022) News: commission acts to better protect people from asbestos and ensure an asbestos-free future (2022). European Commission Site. Employment, Social Affairs & Inclusion. Retrieved 5 Nov 2023, from https://ec.europa.eu/social/main.jsp?langId=en&catId=89&furtherNews=yes&newsId=10418#navItem-relatedDocuments
Feola G (2015) Societal transformation in response to global environmental change: a review of emerging concepts. Ambio 44(5):376–390. https://doi.org/10.1007/s13280-014-0582-z
Folke C (2006) Resilience: the emergence of a perspective for social-ecological systems analyses. Glob Environ Chang 16(3):253–267. https://doi.org/10.1016/j.gloenvcha.2006.04.002
Folke C, Carpenter SR, Walker B, Scheffer M, Chapin T, Rockström J (2010) Resilience thinking: integrating resilience, adaptability, and transformability. Ecol Soc 15(4)
Gopar-Nieto R, Cabello-López A, Juárez-Pérez CA, Haro-García LC, Jiménez-Ramírez C, Aguilar-Madrid G (2016) Actualización sobre la epidemiología, fisiopatología, diagnóstico y tratamiento del mesotelioma maligno pleural*. Rev Méd del Inst Mex del Seguro Soc 54(6): 770–776
Gunderson LH, Holling CS (2012) Panarchy: Understanding transformations in human and natural systems. Island Press
González-Arellano S (2022) The future scenarios of cities: an analysis of their institutional construction. Int J Urban Reg Res 46(5):865–873. https://doi.org/10.1111/1468-2427.13103
Hadley B, Rennell T (2013) Asbestos the future risk (D Austin, ed). Iskaboo Publishing Ltd.
Hölscher K, Wittmayer JM, Loorbach D (2018) Transition versus transformation: what's the difference? Environ Innovation Societal Transitions 27:1–3. https://doi.org/10.1016/j.eist.2017.10.007
Irwin T, Kossoff G, Tonkinwise C (2015) Transition design provocation. Des Philos Papers 13(1):3–11
Irwin T, Tonkinwise C, Kossoff G (2020) Transition design: an educational framework for advancing the study and design of sustainable transitions. Cuadernos 105. https://doi.org/10.18682/cdc.vi105.4188
Latour B (1999) On recalling ant. Sociol Rev 47(1_suppl):15–25. https://doi.org/10.1111/j.1467-954X.1999.tb03480.x

Loorbach D, Frantzeskaki N, Avelino F (2017) Sustainability transitions research: transforming science and practice for societal change. Annu Rev Environ Resour 42(1):599–626. https://doi.org/10.1146/annurev-environ-102014-021340

Manzini E, Jégou F (2003) Sustainable everyday: Scenarios of urban life. Triennale di Milano, Milano. Ambiente

Meadows D (2008) Leverage points: places to intervene in a system 1999. The Sustainability Institute, Hartland, Vermont

Miller R (ed) (2018) Transforming the future: anticipation in the 21st century. Routledge

Morales Zaragoza NA, Barraza Gómez EC (2020) El Enfoque emergente del Diseño para la Transición: Acupunturas sociales, identidades y estilos de vida en relación con el problema del agua en la zona patrimonial de Xochimilco. DIS 5: 1–28. https://dis-journal.ibero.mx/index.php/DISJournal/article/view/71

Nussbaum MC (2013) Creating capabilities: the human development approach (First Harvard University Press paperback edition). The Belknap Press of Harvard University Press

O'Brien E (2022) The "Next" effect: When a better future worsens the present. Soc Psychol Personal Sci 13(2): 456–465. https://doi.org/10.1177/19485506211034972

O'Brien K, Sygna L (2013) Responding to climate change: the three spheres of transformation. Proc Conf Transform Changing Clim 16–23

Patterson J, Schulz K, Vervoort J, van der Hel S, Widerberg O, Adler C, Hurlbert M, Anderton K, Sethi M, Barau A (2017) Exploring the governance and politics of transformations towards sustainability. Environ Innovation Soc Transitions 24:1–16. https://doi.org/10.1016/j.eist.2016.09.001

Preiser R, Biggs R, De Vos A, Folke C (2018) Social-ecological systems as complex adaptive systems: organizing principles for advancing research methods and approaches. Ecol Soc 23(4). https://doi.org/10.5751/ES-10558-230446

Reed MS, Vella S, Challies E, De Vente J, Frewer L, Hohenwallner-Ries D, Huber T, Neumann RK, Oughton EA, Sidoli Del Ceno J, Van Delden H (2018) A theory of participation: what makes stakeholder and public engagement in environmental management work? Restor Ecol 26(S1). https://doi.org/10.1111/rec.12541

Steffen W, Persson A, Deutsch L, Zalasiewicz J, Williams M, Richardson K, Crumley C, Crutzen P, Folke C, Gordon L, Molina M, Ramanathan V, Rockström J, Scheffer M, Schellnhuber HJ, Svedin U (2011) The anthropocene: from global change to planetary stewardship. Ambio 40(7):739–761. https://doi.org/10.1007/s13280-011-0185-x

Valadez-Blanco O, Morales-Zaragoza NA, González-García DN (2020) Catalizando cambios sociales y bioculturales sobre el cáncer. Diseño de acciones transdisciplinarias contra la pandemia. Rev de Salud Pública 22(3): 1–10. https://doi.org/10.15446/rsap.v22n3.87147

Walker B, Holling CS, Carpenter SR, Kinzig AP (2004) Resilience, adaptability and transformability in social-ecological systems. Ecol Soc 9(2). https://doi.org/10.5751/ES-00650-090205

Wallis SL, Emmett EA, Hardy R, Casper BB, Blanchon DJ, Testa JR, Menges CW, Gonneau C, Jerolmack DJ, Seiphoori A, Steinhorn G, Berry T-A (2020) Challenging global waste management—bioremediation to detoxify asbestos. Front Environ Sci 8.https://doi.org/10.3389/fenvs.2020.00020

World Health Organization (2015) Chrysotile asbestos. World Health Organization. Retrieved from https://iris.who.int/handle/10665/178803

Zaragoza NAM, Ugarte BM, García CT (2021) Mapeos generativos, herramientas de participación interdisciplinaria para la transición socio-ecológica. Prototipado de iniciativas. Cuad Del Centro de Estud de Diseño y Comun 132. https://doi.org/10.18682/cdc.vi132.4989

Chapter 9
Anticipation in Three Restaurants of Mexico City: Imagining Edible Fungi in Urban Food Futures

Mónica Velázquez Téllez

Abstract This chapter analyzes the forms of anticipation of three mushroom-based food restaurants in Mexico City in the face of the horizon of urban food transitions in the context of the ecosocial crisis. The first part describes the food scenario proposed by international organizations and the importance of food systems for climate change. The second part describes the relationship between cities and food. The third part analyzes how the mushroom is being created as an emerging object from different fields, and its role in food and the Mexican food system. The fourth part deals with the case studies of the three restaurants (Tencüi, La Hongoteca, and Los Loosers). The fifth section discusses the anticipatory capacities of the restaurants and some little-known elements that the relationship between urban systems and wild mushrooms could make possible.

Keywords Food systems anticipation · Mushroom-based gastronomy · Urban food system futures

9.1 Introduction

This chapter will analyze the anticipation forms of three restaurants located in the Cuauhtemoc district of Mexico City whose gastronomic proposal employs mushrooms. The first is Tencüi, the second is La Hongoteca, and the third is Los Loosers. Anticipation occurs when the future is used in action (Poli 2020), which implies that, based on the reading or interpretation of a given context, the system that interprets this context can act or not act in certain behaviors to respond to a probable future, that which comes from the reading that the system itself makes of the context. Based on the fieldwork in the three restaurants mentioned above, we will analyze what forms of future around food are practiced in these restaurants by implementing a mushroom-based cuisine. Although no one has the last word on what will or will

M. Velázquez Téllez (✉)
Universidad Autonoma Metropolitana-Cuajimalpa, Mexico City, Mexico
e-mail: monica.velazquez@cua.uam.mx; monveltz@gmail.com

S. Gonzalez-Arellano and B. Gandlgruber (eds.), *Cities as Anticipatory Systems*, The Urban Book Series, https://doi.org/10.1007/978-3-031-78162-9_9

not happen in the future (Miller 2007), critical reflection on the future and, especially, conscious anticipatory thinking is vital to respond creatively to the challenges posed by the Anthropocene, a time in which the future is coded not only in terms of uncertainty—as it has always been—but also in terms of survival.

Given the dominant narratives about the multiple socio-environmental crises that this time implies, the role of urban food systems is crucial for imagining food futures that respond to a horizon in which multiple socio-environmental issues are looming. Likewise, the figure of the restaurant as a space for innovation and taste formation constitutes a site of transition that can articulate new food paradigms but also lead to the rise in price and gentrification of foods that were not previously part of the major market trends. The case of the three mushroom restaurants will allow us to observe the role of urban food systems in food transitions, as well as to reflect on the emergence of the mushroom as an object of scientific interest and a powerful metaphor that mobilizes imaginaries toward futures based on interspecies collaboration.

9.2 Scenarios of Food System in a Global Crisis Panorama

9.2.1 Scenarios of Food Crisis

International organizations and scientific groups such as EAT, FAO, and the Lancet have constructed a food scenario in which the multiple socio-environmental crises are strengthened. When considering the different elements that make up food systems and the influences they suffer and exert in a hyperconnected world, the anticipation linked to them is a fruitful topic for reflection. Food systems are the set of all factors and environmental elements, human inputs, processes, infrastructures, and institutions around food (Graziano da Silva et al. 2021). They are a crucial node in which elements of very different kinds intersect since the food field, species, geographies, climates, traditions, technological developments, symbols, governments, and economic powers, among many others are related. In a globalized planet, these relationships, with their asymmetries of power and cascading effects, are even stronger; as an example of this, the pandemic caused by SARS-CoV-2 that started at the end of 2019, probably in the province of Wuhan (Zapatero Gaviria and Barba Martin 2023), shows that there is no small problem when the factors of a local food system are disrupted (Nchanji and Lutomia 2021).

Recently, food has also become a lifestyle (Si 2020; Sexton et al. 2022; Vargas-Lopez et al. 2022) and even a political ecology attitude to embrace (Anguelovski 2015; Hebinck et al. 2018; Pasquier Merino et al. 2022) stance, especially as it relates to environmental justice (Kortetmäki and Oksanen 2021; Sexton et al. 2022) and the relationship between greenhouse gas emissions and animal protein consumption, particularly beef (Willett and Rockstrom 2018). These behaviors indicate that there is something in the food that is emerging with great power, as the decisions of some consumers whose economy allows them to prioritize consumption decisions, begin to conceive food functions and daily habits related to food as enclaves to promote transitions toward more sustainable, healthier or even, for some, more ethical food. The relationship between food choice and promoting some change is a novel behavior in food consumption. It is important to note that each of these food choices responds to an implicit idea of the future so that one decides to eat in one way or another, thinking, for example, that animal welfare will be achieved or that plant-based foods can help reduce greenhouse gas emissions, or that food should also be medicinal to prevent some ailments. Food futures are active in every food choice or omission, as seen below in the three mushroom restaurants.

We mean the superimposition of several negative factors around food systems by referring to the "food crisis" within the global environmental crisis. The first, related to hunger, is the continued malnutrition caused by deficient caloric intake. The second factor is the lack or low intake of vitamins, minerals, and other essential nutrients (hidden hunger). The third crisis factor has to do with the relationship between nutrition and the development of non-communicable diseases resulting from obesity, such as cancer, diabetes, or cardiovascular problems (Díaz 2021). To these crisis factors linking food and health, which are much more closely related to the weight of the food industry and the large corporations that control the food market, we should also add unexpected events or wild cards, such as the COVID-19 pandemic, whose impact on food systems has not yet been fully calculated (Adelodun et al. 2021).

Another crucial element involved in the crisis of food systems is the nexus between the production of space and the form of cultivation. Historically, food systems have produced diverse socio-ecosystems; today, extensive agriculture and monoculture contribute to reaffirming the physical and symbolic distance between the countryside and the city, spaces of production and consumption, producing agents and consuming agents, physical labor, and intellectual labor. Monoculture, allied with the global expansion of cities, produces a space that hinders food sovereignty and security in urban centers while conceiving cities, at least megacities, as spaces disconnected from the life-supporting elements that sustain global food. Food futures are, at least from a narrative, totally related to the socio-environmental crisis; therefore, food and the change in the human relationship with it is imagined as a space of incidence to shape, even the physical characteristics of the biosphere.

It is estimated that by 2050, the transformation to healthy diets will require substantial changes. Global consumption of fruits, vegetables, nuts, seeds, and legumes will have to double, while foods such as red meat and sugar must be reduced by more than 50% (Willett and Rockstrom 2018). The relationship between human and planetary health should be emphasized, as food systems are responsible for one-third of global greenhouse gas emissions (Crippa et al. 2015). For these reasons, food systems are crucial for optimizing interspecies life and safeguarding human health. In this scenario elaborated by international organizations assessing world food conditions, the discipline of anticipation becomes relevant in the transitions needed to articulate a radical transformation that allows the regeneration of the land and an effective integration of the socio-ecosystems that provide food or, at least, to visualize probable responses in the face of uncertainty. It is necessary to note that anticipation is not homogeneous since it depends on readings that belong to specific contexts. In this sense, it is important to analyze to which futures the practices that urban systems carry out in the present respond, for example, if they anticipate market trends, or an ecosocial crisis more linked to justice and the distribution of wealth, or a future in which those who produce food and those who consume it are conceived differently.

9.2.2 Cities and Food

Regarding the link between cities and food systems, it is worth mentioning that the urban population consumes 70% of the food grown in the world (Tefft et al. 2017); this implies that cities play a fundamental role in terms of the food production-distribution-consumption-waste relationship since it is expected that by 2050, 68% of the world's population is expected to live in cities, which makes it necessary to consider a scenario in which it would be favorable for the urban–rural continuum to evolve into cities centered on food production, as implied by the potential that Mexico City still has, as will be seen below.

In the urban context, food systems are particularly affected by six factors, which correspond to (1) demographic changes and migrations; (2) the paradox of rising incomes and widening inequality gap; (3) climate change and scarcity; (4) evolving consumption, nutrition, and health preferences; (5) rapid technological change and innovation; and (6) localized development and strong involvement of stakeholders and decision-makers (ibid.). In this sense, food systems in urban contexts represent a double challenge; on the one hand, they are not exempt from getting involved with the interconnection of apparently distant socio-ecosystems, as demonstrated by the 2020 pandemic or the conflict between Russia and Ukraine, because in cities, due to international mobility and tourism connections, local events are enhanced in a fractal way. On the other hand, in addition to the socio-environmental crisis that affects all food systems, in urban contexts, these effects are also combined with others of a social nature that deepen the inequality gap since "urban food insecurity in low-income countries is greater than in rural areas, reaching more than 70% in urban slums. Informal safety nets help boost access to food" (Ibid., p. 23).

In the present century, migrations of environmental refugees to urban areas produce permanent or transitory displacements in cities (Chung et al. 2022). Some 20% of the migrant population lives in the 20 largest cities in the world (Tefft et al. 2017). For urban food systems, voluntary or involuntary migration caused by climate change or other social crises generates demographic pressure that complicates the role of cities in terms of their ability to ensure food security; on the other hand, the migrant presence also influences the variety of urban gastronomies, as food is also a cultural heritage of peoples' identity (Reddy and van Dam 2020). Likewise, the number of megacities and cities (1–10 million residents) is expected to increase. The growth of urban forms as the hegemonic way of inhabiting the planet causes the abandonment of fields as cultivation areas. It makes it necessary to reorganize cities around food systems or, in other words, it is imperative to reimagine cities as spaces for living and food production.

Life on the planet is articulated around food systems, and food production shapes the planet. It is necessary to change extensive agriculture from monocultures to an ecological approach of family farming (Biel 2016). For these reasons, food systems are a key that could contain clues to reorder life in the face of the emergency derived from the global environmental crisis, since food systems encompass all natural resources of food not only for humans but for other species used in human consumption; thus, food systems also weave value chains at local and global scales while producing specific inequalities and designing the space of the habitable; the latter has important implications for the cities of the future in the key of resilience.

9.2.3 Food Systems in Mexico City

Even though Mexico City's productive food system has an agricultural potential of 49% (Fuentes 2023), the city, at least from the existing relationship between producers and government, does not seem to consider the current scenario of the global food crisis to activate its productive capacities. In government strategies, there are multiple impediments to promoting food productivity in the city, such as the absence of differentiated support that considers the characteristics of productivity per municipality, lack of an internal transportation network that links products to the core areas of consumption, pressure from food industries; generational abandonment of agricultural work, among others. That does not imply that, through other mechanisms, no measure is articulated that seeks to respond to the demands of civil society for food security and sovereignty in search of healthy and fair food that benefits local producers and consumers.

Even though urban agriculture in Mexico City is not widespread and that buying, rather than producing, seems the most attractive idea (Capron et al. 2017), it is true that, in the wake of the pandemic, a generational relay was observed of those unemployed grandchildren who returned to activate the productivity of the land that their parents had abandoned, this in response to the needs of local food supply (Fuentes 2023). As has a strong indigenous base in many aspects of food, traditional food systems based on pre-Hispanic methods coexist with urban ideas of food, for example, those executed from places such as urban gardens. These multifunctional spaces fulfill, at the same time, economic, cultural, and social demands with the capacity to promote education regarding the agro-food systems that survive in the city and contribute to urban environmental services that simultaneously provide food security and sovereignty (Alcántara and Larroa 2022).

Urban food systems can be classified into three types (traditional, modern, or emerging and informal), which generally exist in all large cities to varying degrees, as they overlap and share functions. The traditional system is the one that is articulated around the connections between large stores and markets in the city with the production of the countryside and its mediators. The global or emerging food system corresponds to the series of variants and actors around food, such as wholesale sales and elements that guarantee food safety, such as packaging, refrigeration chains, business services around food, and other technologies, such as food labeling, sales to restaurants, and the global interaction of food in cities. Informal food systems mainly supply the food demand of the city's impoverished populations. These types of systems are represented by transactions mostly in cash, and their forms of sale are generally itinerant, or else they are premises or businesses that are not registered with the formality of state institutions. The case that will be addressed here is the study of the modern urban food system, which involves value chains in which restaurants play an essential role, as seen below. Thus, it is possible to state that the three restaurants analyzed in this chapter are part of the modern urban food system in Mexico City.

About the three mushroom-based restaurants, it is crucial to consider the role that restaurants play in cities in shaping gastronomic taste by introducing and creating dishes that often differ from the most established gastronomic traditions or that use ingredients in other manners. Urban consumers of a medium–high socioeconomic stratum drive an emerging segment in restaurants that is characterized by demand for agroecological, non-GMO products with short production, distribution, and consumption chains (Bertran-Vilà et al. 2022; Tefft et al. 2017). Likewise, due to its purchasing power and cultural capital, this sector is in a position to promote the consumption of less industrialized food congruent with nutritional requirements prepared in urban restaurants that provide an environment or aesthetic experience around the conceptualization of food and the act of eating.

9.3 The Mushroom to the Future: Time to Revisit the Edible Past

9.3.1 Why Fungi Now?

The restaurants studied here concentrate the production of their dishes on the figure of the mushroom as a gastronomic protagonist; this idea is inserted in an international logic that emerges around the mushroom. Because mushrooms are today a food, pharmaceutical, and cosmetic trend, they are at the center of several discussions either from the scientific research of the international mycologist community (Van Court et al. 2022), from the rights of consumption of psychoactive substances or home growing and mycotourism clubs (Buntgen et al. 2017), or from the potential of the fungi kingdom to offer not only nutritious but also functional food. Although the relationship between edible and medicinal fungi has existed since ancient times among most of the world's populations (Pérez-Moreno et al. 2021),), still today, it is a kingdom that remains unknown since approximately 5% of an estimated 2.2–3.8 million global species are known to be familiar (Zhou and May 2023).

Research on the incipient development concerning the psychoactive potential of some mushrooms has been banned in the United States since the 1970s (Zuleta

2022). However, consuming mushrooms with these attributes was never extinct and remained low profile, as something marginal. The prohibitionist policy on the psychoactive attributes of some species had a turnaround because, in 2020, the John Hopkins Center for Research on Psychedelics and Consciousness was established. With this background, the reopening of research on the fungi kingdom was institutionalized, at least in the United States. At least since 2011, some American publications and newspapers referred to the "shroom boom" or "mushroom boom," highlighting the presence of mushrooms as the foods that were beginning to move "to the center of the plates" for being umami, that is, having a versatile flavor known in Japanese as "essence of delight" as it is simultaneously sweet, sour, bitter, bitter, and salty (Searles 2011).

With the confinements caused by the pandemic, the mushroom boom grew; Google search term charts show worldwide that, from April to August 2020, there was a peak in the increase in searches for the word "mushroom." The market for mushrooms (button: *Agaricus bisporus*; shiitake: *Lentinula edodes*; oyster mushroom: *Pleurotus ostreatus*) used in food, pharmaceuticals, and cosmetics was valued at $50.3 billion in 2021, and is expected to grow by 9.7% from 2022 to 2030 due to the increasing population demanding a plant-based diet or non-animal protein-rich food options around the world; in that sense, vegan and vegetarian diets are an influential factor for the development of the market around the mushroom. Mushrooms are also considered in the superfood category for their essential nutrients, such as selenium, vitamin D, glutathione, and ergothionein, which contribute to mitigating oxidative stress and preventing chronic diseases such as cancer, heart disease, and dementia (GVR 2023).

In 2020, Giuliana Furci, founder, and director of the Fungi Foundation, together with a team of Argentines, Chileans, and Brazilians, as well as the American mycologist Donald H. Pfister, proposed an initiative for the fungi kingdom, as well as flora and fauna, to be included in the biodiversity conservation goals from 2020 onward. Thus, the initiative of the three Fs (Flora, Fauna, and Funga) in international assemblies such as the International Union for Conservation of Nature (IUCN) and the United Nations Convention on Biological Diversity (CBD) exemplifies the rediscovery of the fungus and the future of conservation as an institutionalized interest in an international network of knowledge. The conservation of fungi not only implies the conservation of the fungi kingdom, but it is also a strategy focused on the conservation of an entire ecosystem since fungi fulfill specific alliance functions that sustain life, for example, "the vital importance of arbuscular mycorrhizal fungi colonizes 80% of the roots of all plants, a symbiosis that helps plants to establish themselves in the soil" (Bragança 2021, p. 403). Conservation based on the recognition of the fungal kingdom makes the fungus emerge as a prominent figure to influence the possibilities of conserving not only species but environments so that through the fungus and its present conservation, the conception of relationships between life in socio-ecosystems is expanded, thereby expanding the imagination of collaborations between species as well.

Around the fungi kingdom, new approaches are being concentrated to build it up again: the fungus as an object of resilience and forest health as a critical piece in the work of biodiversity conservation, as a potential adaptogen for the new psychiatric medicine that sees in the *psilocybes* an opportunity to treat mental conditions of modernity, such as anxiety, depression, obsessive–compulsive disorders or post-traumatic stress. The different actions motivated by the mushroom's potential seem to envision multiple changes in medical sciences, food, and conservation. However, as a metaphor for cooperation at different levels, the mushroom is also beginning to be constructed as a space for rereading the relationships between humans and non-humans. All these events seem to indicate that point to changes in the scientific, industrial, and cultural spheres around which the mushroom is imagined anew, at least in Western culture.

9.3.2 Mushrooms as a Food Alternative

Just as the fungi kingdom has acquired relevant functions to articulate conservation strategies and drive new research agendas, the food field has again looked toward these species. Edible fungi, the fruit of the mycelial fungal body, are beginning to be considered as an alternative to providing food security (Cuesta and Castro-Ríos 2017) in a world with a growing population because their production occupies less space for cultivation, uses forty times less water than beef protein and thirty times less than chicken protein (Kazer et al. 2022). Approximately 200 of the 2300 edible species are known (Vidal 2000). Likewise, options have begun to appear for home cultivation for self-consumption from food leftovers (Hellwig et al. 2023); in this case, mushrooms also represent an option that reduces the waste caused by leftovers since it is possible to reuse them to feed their growth. There are already companies, such as the U.S. company Quorn, that offer packaged foods made from mycoproteins of the *Fusarium venenatum* variety. Similarly, in 2019, the National Institute of Industrial Technology of Argentina and Enye Tech started cultivating the same variety under the local name "kernel" to focus on exploring new protein and food markets (Bär 2019).

The different interests in researching the mushroom, both for its potential as a food low in greenhouse gas emissions and for its high protein content, show that, although this species, in its edible varieties, has always existed in people's diets, it is now emerging on the horizon as a novelty that seems to be seen as a creative response to many current problems. The mushroom is considered an alternative not only for planetary health since its production reduces the carbon footprint and is an important option in the search for alternatives to supply quality proteins that do not come from animals, but also, the mushroom is part of the functional food, which considers food intake not only as a means of energy and supply of essential nutrients for the human body, but also as a nutraceutical source, since it provides elements to combat various diseases, whether exogenous, such as those caused by viruses and bacteria, or endogenous, such as cancer and diabetes, caused by a malfunctioning of the immune system and the combination of various factors, such as the intake of industrialized food, high in sugars and preservatives (Ishara et al. 2022; Shamim et al. 2023).

All these diverse interests and views on the fungal kingdom are enabling the re-emergence of the fungus as a new scientific object, in the sense that the food industry is interested in investigating it as an option for food transitions that propose diets focused on lower animal consumption in alignment with planetary and human health. Likewise, the mushroom shows that the concept of food is moving toward not only nutritious but also nutraceutical foods, which implies the emergence of a new food paradigm. For these reasons, the mushroom is interpreted as an emblem of resilience at different levels, which motivates images of futures centered on collaboration, in which the process of growth and cultivation is not only related to humans but to factors of cooperation between species, between mushrooms, forests, and humans, between what dies and what is born, as a metaphor for the degrading capacities of the mushroom. From this perspective, the form of food supply may have a sustainable, regenerative potential in ecological, social, and economic terms (Stöckelová et al. 2023).

9.3.3 Mushroom Gastronomic Tradition in Mexico

Mexico was the first country in Latin America to begin commercial production of edible mushrooms in 1933 near Texcoco, in the State of Mexico (Martínez-Carrera et al. 2000). The main variety, whose strain comes from abroad, is *agaricus*, also known as mushroom. Since 2016, researchers from the Colegio de Posgraduados in the state of Puebla initiated a project to strengthen the genetic varieties of this fungus by combining them with specimens of Mexican wild mushrooms to improve the emerging microbial agro-food chain of edible mushrooms in the country. Although mushroom production in small laboratories is currently obsolete at the industrial level because there is specificity in the production of seeds, substrates, and product processing, small laboratories are relevant at the restaurant production level, as evidenced by the case of the Tencüi restaurant, which is part of this study.

At the industrial or large-scale production level, the main Mexican states with consolidated production are Estado de México, Puebla, Veracruz, Oaxaca, Querétaro, Guanajuato, and Jalisco. The most cultivated varieties in Mexico are mushrooms (white and brown). However, the country also produces mushrooms (*Pleurotus*), shiitake (*Lentulina edodes*), deer mushroom (*Ramaria araiospora*), lion's mane (*Hericium erinaceus*), chicks (*Amanita caesarea*), maguey mushroom *(Pleurotus opuntiae*), oak mushroom (*Entoloma clypeatum*), and friar's head (*Boletus aereus*). Mexico ranks second in the diversity of wild edible mushrooms (Pérez-Moreno et al. 2021). Wild mushrooms have historically been used as part of the diet of indigenous and rural communities (Molina-Castillo et al. 2023), especially in the central and southern regions of the country. Regarding the use of edible mushrooms in the Mexican diet, it is common to find them in various preparations of up to 168 different dishes depending on the region (ibid.), such as dishes that include corn, chili, beans (linked to Mexican popular gastronomic culture) or in combination with elements that were introduced from the different moments of contact with foreigners, such as ingredients of animal origin, such as beef, pork, chicken, cheese, cream, butter, among others. In popular Mexican gastronomy, mushrooms can be considered the main dish, together with other vegetables, or a side dish stewed with *epazote* and other *quelites*.

Despite the Mexican gastronomic tradition around the mushroom, the innovations promoted by the restaurant sectors create the mushroom as a different object, which, although it uses traditional or indigenous discourses about it, is conceived more as a food associated with gourmet food and health. That also implies the emergence of new problems in a horizon of market and food industry interests, in which mushrooms also acquire a class distinction as an object of consumption, which may jeopardize the access that communities have to forest resources since, often, as happens in Japan with high-priced varieties such as matsutake (*Tricholoma matsutake*), they are collected for sale instead of being consumed internally.

9.4 Anticipation Forms in Three Mushroom-Based Restaurants

9.4.1 Tencüi: A Laboratory for the Kitchen

Tencüi, which means "to connect" in Nahuatl, is a restaurant in the Santa María la Ribera neighborhood in downtown Mexico City, in the Cuauhtémoc district. The area is considered an image of a glorious past of great architectural and historical value due to its museums and elements, such as the Moorish kiosk, designed by Ramón Ibarrola to represent Mexico at the Universal Exposition of 1884–1885 (Ramírez Crespo 2021). Toward the middle of the last decade, with the arrival of new inhabitants, Santa María La Ribera began to be a cultural spot (Vazquez-Medina et al. 2019) with which the consumption of goods and services also suffered readjustments, as an example of this, is the demand for prepared foods that formed a landscape of culinary cosmopolitanism (ibid.).

Tencüi, a gastronomic proposal inspired by mushrooms, is part of this landscape. Creating such a food concept responds to the combination of interior design trends proposed by Ricardo Casas, the cuisine of chefs Mario Espinosa and Rodrigo Morales, and the science of edible mushroom cultivation by Carlos Briones, an ethnomycologist. This restaurant employs, in situ, a mushroom cultivation chamber, mainly used to produce yellow mushroom (*pleurotus citrinopileatus*) and pink mushroom (*pleurotus djamor*), royal trumpet (*pleurotus eryngii*), resishi (*Ganoderma lucidum*), chestnut mushroom (*Pholiota adiposa*) and lion's mane (*Hericium erinaceus*), edible species that form part of the restaurant's dishes, either as an accompaniment to main courses or as part of beverages (Fig. 9.1). The mushroom laboratory or chamber began in April 2022, the year the restaurant opened.

Fig. 9.1 *Cerdo pelón* in mushroom sauce, inside the culture chamber, developing mycelium, restaurant exterior

Although in Mexico City there are restaurants whose products come from urban gardens or that have their cultivation methods for internal supply, Tencüi has incorporated a cultivation chamber to cover 30% of its requirements in fungal ingredients (Briones 2022), this implies a conception of the restaurant that goes beyond a space designed for the preparation and sale of food for the urban consumption of upper-middle class diners. However, adding the cultivation chamber to the restaurant infrastructure modifies the attributions of the restaurant as a space for food preparation. It turns into a place with hybrid attributes in which the functions of food production are shared, but also the spaces of the specialized science of ethnomycology linked to the cultivation of edible mushrooms. Carlos Briones, in charge of the laboratory until 2023, has trained the team of chefs in the kitchen so that they can also manage the laboratory; this implies new attributions regarding the capacities and spaces of the food circuits in restaurants and those involved in the circuit.

The mushroom chamber in the restaurant functions as a way to solve the supply, thus guaranteeing the origin and quality of its inputs while at the same time making it possible to respond to the problem that mushrooms are seasonal, that is, the operation of a restaurant based on wild mushrooms would depend on the rainy season, however, as an anticipatory practice, the small laboratory performs the artificial function of the natural ecosystem that allows the development of various mushrooms. The restaurant is making it possible to respond to its own demands thereby shaping itself as a small space that imagines restaurants as food producers and not only as traders or socializers. Certainly, integrating a laboratory into the restaurant is not a standard feature defining the "restaurant" space. Therefore, Tencüi proposes a new approach to the restaurant structure, in which food production is part of the food socialization chain through preparing various dishes. In addition, the cultivation chamber also has a kind of function as an experience for the diners who can visit it.

In this way, the Tencüi project can be thought of as a restaurant-laboratory that amalgamates economic interests and design and consumption trends through the creation of an innovative gastronomic option, in the sense that it uses various edible mushrooms for the design of main dishes, desserts, and even cocktails, contributing to the creation of a taste and a type of food option for consumption in restaurants linked to mushrooms. As a hybrid system, Tencüi employs self-cultivation of mushrooms thanks to the artificial conditions offered by the chamber equipped to humidify the environment, thus ensuring that the mushrooms will be available even outside the seasonal rainy seasons while at the same time making it possible for part of its inputs not to depend on external suppliers. Tencüi's anticipatory practices imply the consideration of a future in which its primary ingredient is unavailable at all times of the year, so the in-house laboratory anticipates cultivating various edible mushrooms. At the same time, creating a restaurant-laboratory implies an idea of the urban food future in which the food is produced in situ and functions as a space of mediation between specialized knowledge of mycology and gastronomy that becomes its different dishes.

9.4.2 *La Hongoteca: Gastronomy as Means to Divulgation*

Like Tencüi, La Hongoteca opened its doors in 2022, coinciding with the mushroom boom in other parts of the world, especially in the United States. However, the restaurant project became an effort to link the cuisine and the dissemination of the properties of mushrooms in other types of activities, so the restaurant, as such, is temporarily closed. The restaurant concept came from two partners who decided to start a mushroom-based food business in the Roma neighborhood, in the city's center and west of the Chapultepec Forest. The Roma neighborhood, similar to Santa María La Ribera, has had an increase in young population since 1990 (Mondragón Yépez 2015) and, in recent years, especially as a consequence of the pandemic and the possibilities that digital nomads from rich countries have to travel to cheaper countries and rent housing in cosmopolitan neighborhoods, the foreign population has also increased (Mariano-Juárez et al. 2023).

Ana Rosas, partner in charge of design, marketing, and administration of La Hongoteca, points out that she conceived the concept as a place that "stores or where there are several types of mushrooms" (Rosas 2023), hence the origin of the restaurant's name linked to the nouns with the suffix-*teca* in Spanish that serve to describe a place where something is stored or that gathers several species of the same thing: media library, library, discotheque, film library, among others. In this sense, La Hongoteca, as a restaurant, also aspired to build the idea of a place devoted to mushrooms, not only in their nutritional aspects but also in their cultural value, having gastronomy as a gateway. Rosas (2023) refers to the need to revalue mushrooms as socio-ecosystemic elements that, for some reason, were forgotten or are considered almost unknown and whose symbolic relationship for the bulk of the population lies in toxicity associated with death, psychoactive substances, or the caricatured image of pop culture from video games. These mentions show that the mushroom, especially in terms of its wide range as a culinary ingredient, remains unexploited, at least for urban populations whose daily life is independent of forest resources in an obvious way. For Rosas, the mushroom is not found in the popular imagination beyond a mushroom, a mushroom, or, in the case of central Mexico, a huitlacoche.

The emergence of La Hongoteca as a gastronomic project was possible thanks to the reading of several signs of the emerging context around mushrooms from 2018 to 2022, as Rosas, a mushroom aficionado and trained in marketing, knew how to interpret the culture of the fungi kingdom was going through a point of effervescence. The difficulties of introducing a vegetarian-vegan restaurant whose dishes were mushroom-based resulted in the creation of a menu in which the different varieties of mushrooms used had to go hand in hand with familiar flavors: Ensenada taco (gray mushroom-based); croquettes of the gods (huitlacoche croquettes with bechamel sauce); hamburger (fried chicken-style mushroom) (Fig. 9.2).To socialize the mushroom as the base of the restaurant's dishes, La Hongoteca used elements already consolidated in the popular taste of the public.

The Hongoteca assumes that gastronomy is a gateway to making mushrooms known, as the heading of its menu states: "Let's get to know the unknown, let's question and be amazed." Likewise, a legend on the menu announces: "Our store has curiosities, books, adaptogenic mushrooms with extract, powder and capsules, teas, stickers, art, jewelry, preserves, fresh and dried mushrooms." The restaurant also sought to generate a consumer experience focused on the dissemination of the nutritional, medicinal, and cultural attributes of mushrooms. With this intention, it is

Fig. 9.2 Buffalo wings made of mushrooms, Taco Ensenada of gray mushroom and blue corn, mushroom burger, cream of mushroom soup, restaurant exterior

evident the goal was to imagine the restaurant as a space of mediation in which food is a first step to creating knowledge, interest, and markets that offer experiences such as mycotourism or amateur workshops on the cultivation and properties of mushrooms.

Like the case of Tencüi, La Hongoteca was constituted as a hybrid space in which the dissemination of ethnomycological culture was part of the food sales space. Currently, the restaurant space, understood as a place for preparing and selling food on-site, does not have any explicit feature that characterizes it as an area where the public could go to get information on any subject. That is not a minor issue since, through the construction of hybrid spaces that merge the characteristics of an information or teaching center and the characteristics of a restaurant, La Hongoteca updates a specific idea of food futures, the one that conceives that eating is also knowing; that knowledge and scientific dissemination is part of the production, sale, and consumption of food in the restaurant. As an anticipatory practice, La Hongoteca updates a future idea of the restaurant concept in which education is mixed through gastronomy in the same space.

9.4.3 Los Loosers: Imagining the Mushroom as Resistance

Located today in the Roma neighborhood, the Los Loosers project began in 2011 as a ghost kitchen or dark kitchen that offered few options and whose owner, Mariana Blanco, cooked and delivered orders by bicycle. Unlike Tencüi or La Hongoteca, Los Loosers does not present itself as a mushroom-based, vegetarian, or vegan kitchen. However, its gastronomic bet is articulated through plants and mushrooms as a cuisine without animal-based elements. For Blanco, cuisine and stories play an essential role; the restaurant's menu is structured as a comic strip and manifesto in which the chef presents her life mission as a service to animals (Fig. 9.3). Hence, of her cuisine is based on mushrooms and plants, elements with which Blanco shares a life story, in the sense that, coming from a family in Oaxaca, mushrooms were an integral part of her diet long before the shroom boom.

Blanco (2023) considers that the work of a restaurant is collaborative; for her, to be self-sufficient in the supply of its raw materials in the manner of Tencüi would imply working against the mushroom itself, which collaborates in symbiosis to maintain the health of various elements that make up the entire ecosystem. For these reasons, Blanco is aware that the restaurant is indebted to the mushroom farmers (*hongueras*) of Puebla, Oaxaca, and Tlaxcala, who, with their community knowledge about the recognition, selection, and harvesting of mushrooms, form part of his restaurant's dishes. As an anticipatory practice, Blanco uses dehydration and fermentation techniques to preserve mushrooms out of season; for her at Los Loosers, the mushroom season is 365 days. From the owner's perspective, it is still difficult to make dishes without any element of animal origin to be entirely well received by her public because, even though most of her diners are omnivores, none of her dishes use any animal product. Blanco considers this difficulty to be due to a lack of awareness of animals as terrestrial inhabitants with the same rights as humans. In this sense, for the

Fig. 9.3 Menu. Entrance of the restaurant. Mushroom Ramen

chef, mushrooms do represent a way of eating "ahead of their time" because, in her opinion, the conditions are not yet in place for eating plants and mushrooms to be the standard basis of food. Blanco's comments allow us to observe how the scenario of a diet composed mainly of plants and mushrooms is part of the futures activated in the present practices of his restaurant by way of anticipation. Los Loosers participates in a food that is not yet generalized and thus puts into operation a specific idea about the future of food: scenarios in which the consumption of animals disappears.

As mentioned above, mushrooms are in vogue today. However, when Los Loosers began, the restaurant sought to appropriate those practices that were not commonplace and that belonged to trends that were little understood in the country, such as using the bicycle as the primary means of transportation or food based on plants and mushrooms. Los Loosers, from the choice of its name, uses a marginal discourse as a form of resistance. In that sense, Blanco's stories around the kitchen have a symbolic and primordial role in the act of cooking that she performs because, for Blanco, the dishes are created not only from the combination of ingredients but of stories that enclose memories about food that can remember territories, people, or moments. The chef and sole proprietor points out that her restaurant was gateway that allowed plant-based food to enter the city. Today, Los Loosers still does not proclaim itself as a vegetarian or vegan kitchen, but its owner considers the food space to be her way of remaining in the service of animals. From that perspective, her restaurant plays a role in generating spaces of transition toward food forms that do not use animals in the elaboration of their dishes as a way of embodying what for some is considered a more equitable political stance with lives similar to human forms.

9.5 Reflections on the Anticipatory Capacity of Restaurants

9.5.1 Anticipation Forms

Anticipation can be understood through two dimensions that involve, on the one hand, imagining the future and, on the other hand, implementing actions in the present that respond to what has been imagined of that future (See González-Arellano's Chap. 2 for a deeper discussion), i.e., anticipation is the practice of the future that is actualized in the present. In this sense, the three restaurants studied in this chapter have common and differentiated anticipation practices, generating different responses to emerging problems. On the one hand, from some horizons outlined by studies on food prospectives (Willett and Rockstrom 2018), the need to reduce the consumption of animal proteins is glimpsed. In that line, mushroom restaurants are strengthening a food trend by specializing in food produced mostly from mushrooms, an element that is not yet a common or generalized aspect of the food supply. The mushroom restaurants respond to an image of a food future that seeks to reduce the consumption of animal proteins; thus, the three restaurants implement the anticipation in the sense that their food, mainly made from mushrooms and plants, puts into action the idea of a food future more focused on dishes that do not use animals.

For the moment, this type of cuisine remains in areas of the city that concentrate a large influx of tourism and temporary or permanent residents of foreign origin, which favors the opening of these spaces and makes it possible for certain consumers to pay for these services. The use of wild mushroom varieties as main dishes begins to build an urban taste for foods rich in quality proteins that are not of animal origin; thus, in restaurants, aspirations of future food systems with less environmental impact materialize. In this sense, the restaurant, as a space for dish creation, improvisation, and mixing of elements, directs the taste of the upper-middle class consumer, whose carbon footprint is generally higher than that of the world's poor, toward food with a lower environmental impact that can prepare for major food transitions. This transitional approach can also be seen as an anticipatory practice that activates new forms of consumption focused on acquiring lower environmental impact tastes, thereby actively building for the future.

At the same time that restaurants are responding to food futures rich in vegetables and mushrooms, restaurants based on mushroom gastronomy have been able to interpret the economic trends that enable the mushroom boom by beginning to develop a specialized cuisine and, with it, the development of a specific gastronomic taste for mushrooms of multiple varieties, including wild mushrooms. In these restaurants, gastronomy is not only conceived as a space for food creativity but also as a gateway to socializing the virtues of the mushroom, not only as food but also as an essential part of the planet's health due to its regenerative capacities. On the horizon of species conservation, the cities' demand for wild edible mushrooms could generate positive pressure to strengthen the conservation of oak forests near Mexico City that supply part of the inputs of these restaurants. As explained above, including the fungi kingdom and the protection of flora and fauna implies that legal frameworks are built and enabled to strengthen the conservation of all the elements of the socio-ecosystem, which should include local knowledge communities around the management of mushrooms in forests.

The relationship that the restaurants establish with mushrooms has allowed them to become hybrid spaces in which diverse knowledge coincides, such as disseminating mushroom culture beyond its nutritional aspect or training restaurant personnel in handling and cultivating different mushrooms in the laboratory. The three restaurants extend the frontier of what is known by becoming educational spaces, self-cultivation, or personal projects that seek to embody a position in search of animal equity through the preparation of their dishes. These attributes are not a current trend that describes the typical functions of a restaurant; it means that Tencüi, La Hongoteca, and Los Loosers are inhabiting a transitory space that may or may not become a dominant trend in how food is conceived in cities. On a powerful level, the imagination of the mushroom as a being that cares for the forest and is a bridge between life and death by degrading the organic matter of the forest and nourishing the soils to enable life to continue to grow, feeding the social schemas of collaboration between cities and rural communities through how they eat.

9.5.2 Futures About Mushroom-Based Food

Although, as mentioned at the beginning of this chapter, no one has the last word on the future, the growing interest in wild mushrooms may lead to an over-exploitation of the forests, causing the communities that have historically passed on the knowledge of recognizing and collecting species to be displaced. However, the demand for these foods may also become a strategy that aims at ecosystemic conservation, with the mushroom as the center of all forest diversity. These actions open the door to multiple futures around mushrooms that, at this moment of juncture between specialized studies from the fields of ethnomycology, psychiatry, and food potentials around the fungi kingdom, simultaneously enable the path of forest devastation and the rise in the price of the mushroom as a luxury food or the articulation of value chains that include cities and the countryside in a fair way. In order to move toward these realities, it is essential to imagine what future we wish to achieve through the decisions of the mushroom as a new element in urban food systems.

The boom that mushrooms are experiencing in some food markets also implies unexplored areas for long-term nutrition. Although several studies have noted the nutritional and nutraceutical properties of edible mushrooms (Baruah et al. 2023; Gariboldi et al. 2023; Ghosh et al. 2021; Maury et al. 2011), there are also indications that due to their powers as organic matter degraders and their relationship with soils, edible macroscopic fungi can concentrate large amounts of heavy metals that may pose a threat to human health in the long term (De Lucio-Flores et al. 2021; Domínguez-Zúñiga et al. 2021). There are still unexplored areas about the relationship that mushrooms will have in food; for the moment, they emerge as a figure of interest that concentrates on the imaginary of a future food with sustainability and socio-environmental justice features.

9.6 Conclusion

In this chapter, the practices of three mushroom-based gastronomy restaurants were analyzed from the perspective of anticipation. It is possible to conclude that in all cases anticipatory practices articulate a food future in the present, in which the main ingredients are proteins obtained through the intake of mushrooms. Tencüi foresees and solves the problem of the seasonality of mushrooms with its integrated cultivation chamber; at the same time, this element activates a food future in which the food sold in restaurants comes from in situ production. La Hongoteca anticipates the figure of the restaurant as a hybrid space in charge of educating through the dishes; in the restaurant, notions of the food future are activated as an informative link between sellers and consumers, which is enabled through the figure of the restaurant, in this sense, La Hongoteca anticipates to update futures in which the restaurant is also a place to acquire knowledge. Employing the metaphor of the work of mycelium in ecosystems, The Loosers imagines a food future based on small-scale complex networks with local collaborations that empower wild mushroom gathering communities. It can be argued that, albeit in different ways, all three anticipatory strategies contribute to erecting food futures that transition to the exploration of a new taste in which the consumption of wild mushrooms and mushroom products is a critical factor in developing a new food future.

References

Adelodun B, Kareem KY, Kumar P, Kumar V, Choi KS, Yadav KK, Yadav A, El-Denglawey A, Cabral-Pinto M, Son CT, Krishnan S, Khan NA (2021) Understanding the impacts of the COVID-19 pandemic on sustainable agri-food system and agroecosystem decarbonization nexus: a review. J Clean Prod 318(February):128451. https://doi.org/10.1016/j.jclepro.2021.128451

Alcántara N, Larroa RM (2022) La multifuncionalidad de los huertos urbanos en la Ciudad de México. Espiral xxix(83):187–230

Anguelovski I (2015) Alternative food provision conflicts in cities: contesting food privilege, injustice, and whiteness in Jamaica Plain, Boston. Geoforum 58:184–194. https://doi.org/10.1016/j.geoforum.2014.10.014

Bär N (2019) Tecnología alimentaria: desarrollaron un hongo con alto contenido en proteínas. La Nación. https://www.lanacion.com.ar/sociedad/sin-titulo-nid2225441/

Baruah P, Patra A, Barge S, Khan MR, Mukherjee AK (2023) Therapeutic potential of bioactive compounds from edible mushrooms to attenuate SARS-CoV-2 infection and some complications of coronavirus disease (COVID-19). J Fungi 9(9):897. https://doi.org/10.3390/jof9090897

Bertran-Vilà M, Pasquier Merino AG, Villatoro Hernández JG (2022) Food producers in the Peri-Urban area of Mexico City. A study on the linkages between social capital and food sustainability. Sustainability (Switzerland) 14(23):1–20. https://doi.org/10.3390/su142315960

Biel R (2016) Sustainable food systems. The role of city. UCL, London

Blanco M (2023) Personal interview recorded on zoom
Bragança ACH (2021) Include all fungi in biodiversity goals climate lawsuits could protect Brazilian Amazon The hidden Olympic spectator. Sci Mag 373(6553):403–405
Briones C (2022) Personal interview recorded on zoom
Buntgen U, Latorre J, Egli S, Martinez-Peña F (2017) Socio-economic, scientific, and political benefits of mycotourism. Ecosphere 8(7). https://doi.org/10.1002/ecs2.1870
Capron G, González S, Wigle J, Diez AL, Monterrubio A (2017) The urban food system of Mexico City Mexico (Issue 7). Hungry Cities Partnership, Cape Town
Chung J, Buswala B, Keith M, Schwanen T (2022) Climate mobilities into cities: a systematic review of literature from 2011 to 2020. Urban Clim 45:101252. https://doi.org/10.1016/j.uclim.2022.101252
Crippa M, Solazzo E, Guizzardi D, Tubiello FN, Leip A (2015) Anthropogenic GHG emissions. Nat Food. https://doi.org/10.1038/s43016-021-00225-9
Cuesta MC, Castro-Ríos K (2017) Mushrooms as a strategy to reduce food insecurity in Colombia. Nutr Food Sci 47(6):817–828. https://doi.org/10.1108/NFS-03-2017-0039
De Lucio-Flores SA, Otazo-Sánchez EM, Romero-Bautista L, Gaytán-Oyarzún JC (2021) Hongos macroscópicos como bioacumuladores de metales pesados. Pädi Bol Científico Cienc Básicas Ingenierías Del ICBI 8(16):60–65. https://doi.org/10.29057/icbi.v8i16.5823
Díaz E (2021) El rol de la innovación institucional, tecnológica y financiera en la transformación de los sistemas alimentarios en América Latina y el Caribe, en el contexto de la agenda 2030 y de la pandemia de covid-19. In: Graziano da Silva et al (eds) Sistemas alimentarios en América Latina y el Caribe. Desafíos en un escenario pospandemia, FAO, CIDES, Panamá, pp 71–94. https://doi.org/10.4060/cb5441es
Domínguez-Zúñiga LI, Puente-Valenzuela CO, Estrada-Arellano JR, Aguirre-Acosta E, Aguillón-Gutiérrez DR (2021) Concentración de metales pesados en hongos de la zona metropolitana de la Comarca Lagunera, México. Sci Fungorum 52(November 2020):e1389. https://doi.org/10.33885/sf.2021.52.1389
Fuentes M (2023) Lecture on Agroecosistemas: relevancia socioambiental y alimentaria. Planeación diferenciada. La transición del sistema alimentario en la CDMX. https://www.youtube.com/watch?v=Ww2vWiPoIJw&list=PLqGZhNa0qbM1sgzp-k-MmzgG6dhmyEMmW&index=3
Gariboldi MB, Marras E, Ferrario N, Vivona V, Prini P, Vignati F, Perletti G (2023) Anti-cancer potential of edible/medicinal mushrooms in breast cancer. Int J Mol Sci 24(12). https://doi.org/10.3390/ijms241210120
Ghosh S, Nandi S, Banerjee A, Sarkar S, Chakraborty N, Acharya K (2021) Prospecting medicinal properties of Lion's mane mushroom. J Food Biochem 45(8):1–13. https://doi.org/10.1111/jfbc.13833
Graziano da Silva J, Jales M, Rapallo R (2021) Sistemas alimentarios en América Latina y el Caribe—Desafíos en un escenario pospandemia. Panamá, FAO y CIDES.https://doi.org/10.4060/cb5441es
GVR (2023) Market analysis report. Mushroom market size. 2022–2030. ReportLinker, Lyon
Hebinck A, Vervoort JM, Hebinck P, Rutting L, Galli F (2018) Imagining transformative futures: participatory foresight for food systems. Special feature on designing transformative spaces for sustainability in social-ecological systems. Ecol Soc 23(2):19
Hellwig C, Bolton K, Häggblom-Kronlöf G, Rousta K (2023) Aspects affecting food choice in daily life as well as drivers and barriers to engagement with Fungi-based food—a qualitative perspective. Sustainability (Switzerland) 15(1). https://doi.org/10.3390/su15010317
Ishara J, Buzera A, Mushagalusa GN, Hammam ARA, Munga J, Karanja P, Kinyuru J (2022) Nutraceutical potential of mushroom bioactive metabolites and their food functionality. J Food Biochem 46(1):1–31. https://doi.org/10.1111/jfbc.14025
Kazer J, Orfanos G, Gallop C (2022) Quorn footprint comparison report. Carbon Trust, United Kindom, pp 1–33
Kortetmäki T, Oksanen M (2021) Is there a convincing case for climate veganism? Agric Hum Values 38(3):729–740. https://doi.org/10.1007/s10460-020-10182-x

Mariano-Juárez L, Medina FX, Conde-Caballero D (2023) Popular cuisines at foodie tables. Social class, distinction, and food transformation and incorporation in contemporary gastronomies. Int J Gastron Food Sci 33:100791. https://doi.org/10.1016/j.ijgfs.2023.100791

Martínez-Carrera D, Larqué-Saavedra A, Aliphat M, Aguilar A, Bonilla M, Martínez W (2000) La Biotecnología De Hongos Comestibles en la seguridad y soberanía alimentaria de México. Puebla, II Foro Nacional sobre Seguridad y Soberanía Alimentaria, AcademiaMexicana deCiencias-CONACYT *Y Seguridad Alimentaria.* Pp 1–12

Maury GL, Humberto Joaquín Morris Quevedo D, Albear JM, Chibás LC, Rosa Catalina Bermúdez Savón D (2011) Plantas y hongos comestibles en la modulación del sistema inmune. Rev Cubana Inv Biomedicas 30(4):511–527

Miller R (2007) Futures literacy: a hybrid strategic scenario method. Futures 39(4):341–362. https://doi.org/10.1016/j.futures.2006.12.001

Molina-Castillo S, Espinoza-Ortega A, Thomé-Ortiz H, Moctezuma-Pérez S (2023) Gastronomic diversity of wild edible mushrooms in the Mexican cuisine. Int J Gastronomy Food Sci 31(September 2022). https://doi.org/10.1016/j.ijgfs.2022.100652

Mondragón Yépez M (2015) Proceso de gentrificación de la Colonia Roma.pdf. Universidad Nacional Autónoma de México

Nchanji EB, Lutomia CK (2021) COVID-19 challenges to sustainable food production and consumption: future lessons for food systems in eastern and southern Africa from a gender lens. Sustain Prod Consumption 27:2208–2220. https://doi.org/10.1016/j.spc.2021.05.016

Pasquier Merino AG, Torres Salcido G, Monachon DS, Villatoro Hernández JG (2022) Alternative food networks, social capital, and public policy in Mexico City. Sustainability (Switzerland) 14(23). https://doi.org/10.3390/su142316278

Pérez-Moreno J, Guerin-Laguette A, Rinaldi AC, Yu F, Verbeken A, Hernández-Santiago F, Martínez-Reyes M (2021) Edible mycorrhizal fungi of the world: what is their role in forest sustainability, food security, biocultural conservation and climate change? Plants People Planet 3(5):471–490. https://doi.org/10.1002/ppp3.10199

Poli R (2020) Handbook of anticipation. https://doi.org/10.1007/978-3-319-31737-3

Ramírez Crespo FS (2021) Santa María la Ribera : gentrificación y producción de fronteras sociales. El Colegio de México

Reddy G, van Dam RM (2020) Food, culture, and identity in multicultural societies: Insights from Singapore. Appetite 149(September 2019):104633. https://doi.org/10.1016/j.appet.2020.104633

Rosas A (2023) Personal interview recorded on zoom

Searles BR (2011) A boom in ' Shrooms. Psychology Today, New York, pp 45–46

Sexton AE, Garnett T, Lorimer J (2022) Vegan food geographies and the rise of Big Veganism. Prog Hum Geogr 46(2):605–628. https://doi.org/10.1177/03091325211051021

Shamim MZ, Mishra AK, Kausar T, Mahanta S, Sarma B, Kumar V, Mishra PK, Panda J, Baek KH, Mohanta YK (2023) Exploring edible mushrooms for diabetes: unveiling their role in prevention and treatment. Molecules 28(6). https://doi.org/10.3390/molecules28062837

Si LW (2020) Trending foods and beverages.Food Soc 305–321.https://doi.org/10.1016/b978-0-12-811808-5.00016-7

Stöckelová T, Senft L, Kolářová K (2023) Sympoietic growth: living and producing with fungi in times of ecological distress. Agric Hum Values 40(1):359–371. https://doi.org/10.1007/s10460-022-10366-7

Tefft J, Jonasova M, Adjao R, Morgan A (2017) Food systems for an urbanizing world.https://doi.org/10.1596/32502

Van Court RC, Wiseman MS, Meyer KW, Ballhorn DJ, Amses KR, Slot JC, Dentinger BTM, Garibay-Orijel R, Uehling JK (2022) Diversity, biology, and history of psilocybin-containing fungi: suggestions for research and technological development. Fungal Biol 126(4):308–319. https://doi.org/10.1016/j.funbio.2022.01.003

Vargas-Lopez A, Cicatiello C, Principato L, Secondi L (2022) Consumer expenditure, elasticity and value of food waste: a quadratic almost ideal demand system for evaluating changes in

Mexico during COVID-19. Socio-Econ Planning Sci 82(PA):101065. https://doi.org/10.1016/j.seps.2021.101065

Vazquez-Medina JA, Hernandez Cordero A, Lagos SI, Rangel Trujullo FR (2019) Gentrificacion y alimentacion en Santa Maria La Ribera. Estud Sociales. https://www.scielo.org.mx/pdf/esracdr/v30n55/2395-9169-esracdr-30-55-e20838.pdf

Vidal DR (2000) Ciencia y tecnología de alimentos. Arbor 166(653):127–139. https://doi.org/10.3989/arbor.2000.i653.1004

Willett W, Rockstrom J (2018) Alimentos Planeta Salud. La Comisión EAT-Lancet 32

Zapatero Gaviria A, Barba Martin R (2023) What do we know about the origin of COVID-19 three years later? Rev Clin Esp 223(4):240–243. https://doi.org/10.1016/j.rce.2023.02.002

Zhou LW, May TW (2023) Fungal taxonomy: current status and research agendas for the interdisciplinary and globalisation era. Mycology 14(1):52–59. https://doi.org/10.1080/21501203.2022.2103194

Zuleta P (2022) Interview about: uso terapéutico de los hongos alucinógenos: "todavía tenemos mucho por recorrer". France 24. https://www.france24.com/es/programas/salud/20220826-uso-terap%C3%A9utico-de-los-hongos-alucin%C3%B3genos-todav%C3%ADa-tenemos-mucho-por-recorrer

Chapter 10
Anticipatory Systems and Security in Mexican Cities. A Case-Based Review

José Luis González Madrigal

Abstract This chapter aims to present a reflection on some examples of urban security anticipation. Some attempts at anticipation based on urban actors are reviewed, as well as an analysis of their influence on the anticipatory capacity of the city in general. This will allow us to address the central question of the analysis: can a city by itself constitute a security anticipation system? In terms of security, it could be considered that one of the main problems in anticipating the future in cities is the existence of actors claiming both security and insecurity conditions. The discourses, actions, and mechanisms that allow individuals to anticipate a situation of insecurity are confronted with opposing intentions, making anticipation a difficult process. The chapter recovers concepts of anticipation, such as future literacy, identifying their relevance within an anticipatory system. From there, two cases of Mexican cities are taken up, which allow us to review these ideas in terms of security and anticipation. This discussion allows an analysis of the different types of urban actors involved in security, as well as an approach to their anticipation systems. Finally, it concludes with some considerations about security anticipation at the urban level.

Keywords Anticipatory systems · Urban security · Cities · Urban studies · Anticipation · Futures

10.1 Introduction

To speak of cities today is to think not only in terms of the space they occupy, the inhabitants who occupy them or the socio-economic dynamics they present, but also in terms of the potential that cities represent for individuals. On the one hand, security is an inalienable promise of the city, as well as one of the most challenging urban conditions. Between the search for security and the existence of a market of insecurity, the anticipation of security in cities is in constant dispute.

J. L. González Madrigal (✉)
Universidad Autonoma Metropolitana, Mexico City, Mexico
e-mail: jose.gonzalez@cua.uam.mx

S. Gonzalez-Arellano and B. Gandlgruber (eds.), *Cities as Anticipatory Systems*, The Urban Book Series, https://doi.org/10.1007/978-3-031-78162-9_10

This chapter aims to approach a reflection on the urban capacity to anticipate in terms of security. To achieve the above objective, the first section develops the relationship between cities, their aspiration for security and their need for anticipatory capacities. To do this, it draws on concepts such as future literacy, as well as its relevance within an anticipatory system. In general, some elements that allow the individual to anticipate are identified, in addition to linking this with elements related to security, in such a way that it is possible to approximate what is the basic structure for an anticipatory system in terms of security.

In a second section, we present the cases of two Mexican cities where an exercise of anticipation is intended, based on the experience of proposals and interventions for public security. The first corresponds to a deliberate exercise on the part of security agencies to break with the current situation of insecurity for the inhabitants of Juárez, Chihuahua: the *Todos Somos Juárez* program. The other case corresponds to a coordinated exercise between state agencies and citizens, in which the development of more desirable futures, different from what the prospective marked: the project Red *Conéctate por tu Seguridad*, of the Azcapotzalco district in Mexico City. From this, a third section shows an analysis of the anticipatory capacity for different actors, relevant to a security scheme, will be developed. The first of these is that of security promoters, among whom the governmental structure and its supporters are identified. The second group corresponds to the beneficiaries of security services, including citizens who have no authority to provide or reduce security. Finally, detractors of security, who benefit from breaking the law and violating others, as well as from the deterioration of the security situation, such as those who are not in favor of the governmental structure, the promoters of insecurity. The main delimitation criterion for these groups is the intention to change the security situation in the future.

The document concludes with an analysis of the state of literacy and the future in terms of security that present the cases analyzed. It is based on the above that the document aims to reflect on how the interaction of these different perspectives and aspirations in terms of security contributes to the possibility of integrating an urban anticipatory system. The question that will guide the discussion in the following: how do cities form an anticipatory system in terms of security?

10.2 About Cities, Security and Anticipatory Systems

When we talk about cities, we are not just talking about a community of people with common goals. Nor is it just the network of buildings and infrastructure that allows industries and services to operate. The dynamics and influence of cities go beyond these delimited concepts and are also interpreted as an instrument of human use; the urban has been consolidated as an environment that contributes to transcendental transformations that affect the environment and individuals (Brenner 2013).

In addition, a rhetoric of promise, of a shared desirable future, is usually introduced into the discussion when talking about cities. Among the most common expectations attributed to an urban society are ideas of development, quality of life, economic growth, among others. In general, it can be seen that for each of these promises, the underlying idea has been modified in order to add desirable elements, always with a view to what a city can be and what a city will be. This implies an important complexity about what we can expect from urban life.

In this sense, cities are always in a state of comparison between what they are and what they can be for their citizens. Their decision-makers and inhabitants have configured an expected future within their routine functional logic. This future serves as a guide for the design and operation of cities. In this way, the exercises of anticipation have specialized in the collection and analysis of statistics and metrics, in the creation of thematic observatories, as well as in the exercise of strategic planning in all areas that concern cities: environmental, economic, labor, among others.

10.2.1 An Uncertain Future: Urban Insecurity

One of the promises whose future is more difficult to predict relates to an essential requirement for living in cities: security. As a desirable ideal, security has been one of the motivations for the concentration of a large part of the population in large urban settlements. However, contemporary societies have seen that security remains a promise that seems out of reach, at least in the short term. Moreover, social expressions of insecurity seem to be increasing significantly, not only in frequency, but also in the nature and vulnerability they represent for the inhabitants.

The phenomena of (in)security have changed and become more complex as the urban dimension has grown. This is due to the increasingly specific demands that societies are placing on their social constructs. At the beginning of the discussion, the concept of security referred only to national security, which sought to protect territory from threats between states (PNUD 1994). A second approach is public security, which is the instrumentalization of government policies of territorial control based on the preservation of order in society (Montero 2013). In an alternative sense, internal security rejects the premise that it is a country, nation or territory that needs to be defended or from whom it needs to be defended; it is society that deserves protection and its dimensions of vulnerability are diverse (Montero 2013).

According to Krauze (1998), security approaches began to broaden to include new elements of threat and to be grouped into three main analytical segments. First, those that seek to protect a nation within state boundaries. A second approach remained within the spectrum of state action, but focused on the strengthening of inter-institutional and inter-state arrangements. Finally, those that broaden the protection agenda, taking the protagonism away from the state by including more individual issues, but extending the boundaries of attention to regional and global spaces.

The complexity of the notion of security takes into account not only differences in the criteria for government action, but also the objective it pursues. In this sense, human security emerges as an alternative to the notion of risk reduction and proposes as an additional objective that of peaceful coexistence in a multicultural world (Montero 2013). The duality of human security thus corresponds to the search for peace and development (PNUD 1994). According to Carreón and García (2014), the first approach corresponds to the solid modernity approach, which involves competition for political, territorial, legal, rational, bureaucratic, productive and authoritarian legitimacy, while the second type of approach represents the society of liquid modernity, which has transformed the legitimizing discourse through skepticism, which has brought risk, uncertainty, chance, and insecurity to the social individual.

The diversity of approaches from which the notion of security has been adapted points to a transformation of what is considered threatening. The individual's perception of his or her social and physical environment is what he or she identifies as an enemy or threat. This is particularly relevant in increasingly diverse societies, since the greater the complexity and size, the greater the uncertainty about the different environments and other inhabitants.

An example that synthesizes the above can be seen in the General Security Strategy in Mexico. In it, the Gobierno de México (2018) proposes various actions to achieve security, including the reorganization of police structures, operational guidelines, and the redesign of the functions of federal, state, and municipal agencies. However, the document also recognizes actions aimed at strengthening the ethical perspective of citizens, regaining the trust of citizens, formulating a culture of peace, among other actions that include issues not directly related to crime and violence.

In this sense, it is possible to recognize that the threat has approached the individual from various fields of action. When an individual or group does not behave like the rest of the population (due to lack of incentives to accept the institutional social arrangement), they will seek and provoke the disruption of the system (Foucault 2006), becoming a foreigner (risky) in their own territory. In this way, violence has become increasingly close to the immediate environment of the individual, forcing him to understand his counterpart no longer as an outsider, but as part of his environment; the disruption of peace is made by the neighbor, the community itself (Bauman and Donskis 2015).

It is also important to consider that the complexity of these risks is not absent from the intentionality of some individuals. The idea of criminal and violent acts as an economic instrument has favored the creation and maintenance of criminal groups that, organized or not, find profit in maintaining a climate of insecurity. This is particularly relevant at a time when criminal groups are professionalizing the work of insecurity in order to increase profits and compete for the power that violence brings.

On the basis of the above, it can be said that urban insecurity has become a complex construction in analytical terms, as well as in terms of anticipation. If the threat has diversified in terms of its causes, in addition to the fact that its proximity to the individual has increased, people living in cities find it more difficult to anticipate in a real way the risk to which they are exposed. In addition, the constant intervention of individuals or groups that seek to perpetuate crime and violence in the city creates a crossroads of security strategies.

10.2.2 Security and Anticipatory Systems

It is a fact that security is experienced differently in each city, but in general they show increasing rates of insecurity. This condition is particularly relevant at the Latin American level, since these are the cities that present the highest risk situation (Muggah et al. 2016). However, despite the complexity that cities present to understand and anticipate risk and threat conditions, the expectation generated by the city itself forces to rethink and reformulate new ways of approaching security. Security organizations are forced to broaden their scope in terms of policy areas and expand their strategic capacity to also anticipate developments and risks (Habegger 2022).

This section recognizes the existence of multiple efforts by city agencies to technify their work to provide security in cities. The mythification of technology and the so-called smart cities has generated an interesting debate on the use of technology for security actions. Among many alternatives, information technologies, cybersecurity and the Internet of Things have intervened in security processes (Vergel-Clavijo and Guerrero-Bayona 2023; Lechner 2016).

However, even though the proposals pretend to prevent, control, and deter the criminal acts, some of the results are not as conclusive as expected. Some of the initiatives only represent a way to technologize the security department, without representing an effective change in the security situation or an effective use of this technology (Alvarado 2018). In other cases, it's possible to observe that the initiative itself has brought unwanted risks (Williams et al. 2016).

In terms of the future, those technology initiatives are focused on rather reactive aspects, almost all of which are designed to combat violence, focusing on the effects of insecurity rather than providing security conditions for citizens. Even for those alternatives that could be interpreted as disruptive proposals, there is no indication of what kind of radical change or paradigmatic shift is being sought (Csernatoni and Oliveira 2023). In this sense, another approach that has taken place when talking about security and the use of futures has to do with thinking not in terms of a prospective logic that prevents insecurity, but in terms of resilience and adaptation to imminent risk phenomena (Habegger 2022).

Talking about the prospective of security in cities is not a new topic, since for several decades there has been evidence of cities that have developed initiatives to improve security conditions based on two key considerations: time projection and an orientation of the strategy to address risk (Muggah et al. 2016). But, as discussed in Chap. 1 of this book by González, talking about cities and futures requires both understanding them from the complexity of their system, as well as thinking about the mechanisms through which cities implement visions that have not yet materialized and transform them into processes of action. For this, it is possible to rely on the concepts of anticipation systems and future literacy.

With regard to anticipatory systems, these are tools (technological, social, cognitive, etc.) that allow actors to think about the future and act in the present, based on what the system allows them to observe. For Aquilué and Ruiz (2021), the anticipation of a system corresponds to the capacity to anticipate changes in the environment; to reduce uncertainty, understood as the complexity of the environment.

In a system such as the city, this logic corresponds to deliberate exercises of anticipation by the urban collective, subsystems that are formed to read, interpret, and give meaning to environmental, social, and economic dynamics in order to anticipate (a) what is likely to occur or (b) what may occur based on certain behaviors. It is from this reading of what the city will be in the future that efforts are proposed within the city to access or, as the case may be, move away from that possibility.

As Poli (2019) recognizes, anticipation occurs when the future is used for action. For the author, the above implies a fundamental notion of anticipation, but also the capabilities to take advantage of future information, starting from the intervention of a present situation. In this sense, cities use anticipation exercises with different levels of complexity, ranging from foresight exercises to trend analysis. Cities that present successful security interventions are those that are able to establish strategies thought in short and long-term horizons (Muggah et al. 2016).

According to Boer et al. (2018), it is possible to articulate capabilities, making sense of future literacy. This concept can be disaggregated from an analytical perspective into three considerations: the articulation of anticipatory systems; the use and exploitation of futures (contingent, optimizing, and exploratory); the construction of a sense of use for the information resulting from anticipatory systems.

González-Arellano (2022) considers that the construction of futures is a regular, cumulative, and collective process, since they are the result of an interactive process of narratives. This construction corresponds to the identification of a possible imaginary for those who think about it. If this is true, then the capacity of cities to imagine futures depends directly not only on the outcome of the interaction between individuals, but also on the capacity of its members, including their biases or intentions, in the process of producing future imaginaries.

More generally, this chapter argues that the anticipatory capacity of cities in relation to security cannot be interpreted in a comprehensive way for a city as a whole. It is noted that some cities propose anticipation as a decision-making tool, but it is also observed that there is more than one competing anticipation system, which together inhibit the proper implementation of each one. In this sense, Aguirre (2016) points out that the anticipatory capacity of state agents in security matters is insufficient as long as they are unable to anticipate the operation of criminal agencies.

In this sense, an urban security anticipation system is understood as one that allows, in a clear and simple way, (1) the future literacy (of its actors) on security, (2) to translate this ability to read (foresee) possible scenarios into actions in the present for the transformation of some of the causes that trigger the phenomena of insecurity, as well as (3) to give a sense of use to the information of the future. In this sense, it is not enough to take measures to prevent crime; it is necessary to implement actions that promote a climate of security.

From the perspective of anticipation proposed in this section, it could be argued that one of the main problems in developing anticipatory systems in relation to urban security is the existence of actors who seek security, but also those who prefer insecurity. In this sense, the discourses, actions and mechanisms that allow individuals to anticipate an insecure situation are confronted with opposing intentions and actions, thus hindering the process.

10.3 Urban Security in México. Examples of Futures-Based Action

In addition to the complexity of understanding what insecurity is, the identification of causes and motivations are diverse in the way they manifest themselves and in the social sphere to which they belong. Some of the factors that foster a climate of insecurity are poverty, precarious infrastructure, domestic violence, youth gangs, impunity in the administration of justice, low social cohesion, presence of weapons and drugs, corruption in government agencies, poor education system, child abuse, among others (Arriagada 2001; Gobierno de México 2021; Herrera-Lasso 2013). In this sense, the way in which insecurity manifests itself in the city varies as much as the factors that cause it are present in the city.

In particular, the scope and intentions of urban security schemes vary from city to city. From a forward-looking perspective, crime prevention as an approach is one of the simplest mechanisms used by local governments. It corresponds to a strategy that anticipates an outcome and integrates actions in the present to either promote or prevent it. Since the 1990s, the formulation of strategies that accompany the fight against crime with prevention programs has had favorable results in Latin American cities, compared to those that seek only to combat crime (Arriagada and Godoy 1999).

This chapter seeks to deepen the analysis of the use of futures for urban security, in order to contribute to the reflection on anticipation. In this sense, the next section examines some proposals that, in addition to prevention approaches, recognize the review of future scenarios, and transform them into present decisions that seek to modify the observed trend in a much more active way. Each of the examples will briefly develop what the program consisted of and then review it from the perspective of anticipation.

In order to provide evidence for the central discussion of the chapter, one case was selected in which the exercise corresponds primarily to the governmental sphere, and the second case corresponds to a civil society effort. The purpose of bringing these perspectives together is to provide information about different positions and the construction of futures by different urban actors.

10.3.1 Enough of This Present: Governmental Program "Todos Somos Juárez"

As a government response to the significant increase in violence and crime, the *Todos Somos Juárez* (TSJ) program was implemented in Ciudad Juárez (Juárez City), in the state of Chihuahua, México. Although the program was short-lived (the efforts were concentrated mainly in 2010 and 2011), there was a significant decrease in the number of manifestations of insecurity, as well as an increase in citizen intervention in these matters. It even went so far as to try to replicate it in other cities, such as Acapulco (Rosas 2012), and to use similar mechanisms in other Mexican cities (Arratia 2017).

Although the program reacted to the critical conditions of insecurity experienced in the city, it structured a redesign of security policies in a transversal way, looking for the present causes of a future problem (increase or maintenance of the level of violence). This made it possible to outline preventive measures to deal with the problem, but also to implement emergency measures to deal with the problems experienced. In general, the structure of the effort was intended to encompass the three levels of government, to work intersectorally, to operate intensively, and to actively involve citizens in the diagnosis of the problem and in the operation of the program (Muggah et al. 2016).

A fundamental part of the program consisted in the creation of working groups to identify alternative solutions to the problem of insecurity through intensive citizen participation (op. cit.). Based on the intention of restoring order and the social fabric, resources were used for the implementation of 160 actions in six areas: employment, social development, education, economy, health, and public security (Paniagua and Camargo-González 2017).

In terms of anticipation, the operation of the program highlights the integration of working tables with the purpose of (a) diagnosing the problem (present situation) and (b) establishing routes for action (visions of the future). In addition, the intensive intervention of resources, objectives, and instruments on the city suggests the intention to bring to a convulsive present some tranquility of what the program could offer in the future.

According to Arratia (2017), the results obtained by *Todos Somos Juárez* have been beneficial for the city in two ways. The first is related to the improvement of conditions that allow social development: investments in education, health, economic incentives such as social assistance and credit, among other actions. In terms of security, there has been a decrease in six types of crime: homicide, car theft with and without violence, kidnapping, extortion, and commercial robbery.

Some of the deliberations on the operation of *Todos Somos Juárez* are coinciding with an effective application in the short term, but failed in its execution in the medium and long term. It is even said that once the program was implemented, the longer-term horizon was forgotten (Muggah et al. 2016). Some analysts attribute this trend simply to the political and governmental changes that occurred due to the presidential transition that took place in 2012. However, there is evidence of security effects that were not intended by the policy, one of which was an increase in cases of violent acts by local security agencies (Arratia 2017).

According to Sumano (2018), violence and crime are still manifested in Juarez after the results of the *Todos Somos Juarez* operation, but what has changed is the way they are expressed in the city. For example, the reduction of extortion and rent-seeking crimes in the city is acknowledged; however, the number of homicides, while decreasing in the short term, increased again after the operation. In addition, the city has seen an adjustment in criminal strategy, as there has been a change not only in the cartel present in the area, but also in the drugs consumed by a significant portion of the population, as well as in the type of crime committed in the city.

10.3.2 Proposing Community-Based Futures: Citizen Empowerment Projects and "Red Conéctate por Tu Seguridad"

Another example of a response supported by the review of the futures can be seen in the citizen empowerment projects of the civil association *Mexicanos Unidos Contra la Delincuencia* (MUCD), and more specifically in the *Red conéctate por tu seguridad* (RCS) project, implemented in different municipalities in Estado de México state. The duration of these projects is about 3 years, with most of the actions identified between 2019 and 2021. Projects of similar intent have been carried out in other municipalities of the same state.

Based on the initiative of MUCD, a series of projects have been implemented in the years around 2019 to strengthen social cohesion through the promotion of citizen competences, improve community coexistence, and promote the recovery of public space (MUCD 2019a, b; Alcaldía Azcapotzalco and MUCD 2021). Although the objective of these programs is directly expressed as improving citizen security, since they do not operate on the basis of government action, the focus was not on fighting crime, but on transforming the conditions in which citizens operate.

The general methodology from which the program operated was the training of "facilitators" who would enter into a dynamic of training more trainers and creating networks of citizens to (a) increase social cohesion by locating and strengthening leadership in the community, as well as (b) activate citizens in the implementation of collective actions and citizen empowerment. The idea is the operation of actions from the citizenry, as well as the strengthening of a citizen criterion of respect and care for the environment and the conditions of legality (MUCD 2019a).

The results of the programs do not focus on the actual reduction of crime, but on the transformation of social dynamics. Relevant for the present reflection is the replication of the working groups, not by MUCD, but by the citizens who participated in the primary sessions of the project. That is, the adoption of alternative methodologies to understand and confront the problems present in the community, as well as the proposal and execution of projects by citizens.

In terms of anticipation, the program did not operate with the search for or identification of a specific future, but rather favored the formation of citizens' capacities to visualize and differentiate between possible futures based on their interventions in the community. According to the results reports, the program seeks to train citizens who can lead their community to an environment of legality, not by sharing specific topics and information, following a rigid scheme of citizen training, but by identifying, strengthening, and fostering citizen capacities.

The long-term goals of the project are to achieve peaceful coexistence, the prevention of violence and crime, and the orderly and peaceful use of public spaces (op. cit.). This objective is set not only without a clear roadmap, but also without specific attributions, as could be observed in projects implemented by the government. The bet corresponds to the formation of networks that are adaptive and proactive, based on projects for the community itself.

The identification of a future to be achieved and the recognition of a limited capacity for action in the medium and long term led to the integration of the project into a series of subprojects which, although not articulated, correspond to proposals for present actions for this desirable future. They respond to a kind of "echo" from which present actions modify the state of the question and, in the following deliberation, the network will be able to integrate new action alternatives for this new present, based on the new future identified.

In the specific case of the municipality of Azcapotzalco, the project was carried out in coordination with the mayor's office, which has made it possible to link the citizen's exercise with government action. The goal of this coordination is not only to build community networks, but also to take advantage of other mechanisms for monitoring insecurity by the security agencies in order to improve the understanding of the criminal phenomenon (Vistazo al futuro 2021). For this part of the project, no open evidence of results is identified due to the modifications or adjustments made to the task of combating insecurity.

10.4 Security Anticipation Capacity

As mentioned above, the anticipatory capacity of a city is usually interpreted as the prevention and foresight mechanisms and tools developed by security agencies. However, it has also been argued that this usually does not effectively refer to the anticipatory capacity of the city itself, since it is possible to identify actors who intervene in the security situation, making the anticipation process difficult.

In order to analyze these capacities on the basis of the anticipation literature, it is necessary to recall that anticipatory capacity can be summarized as that literacy about the future of security that allows foreseeing possible scenarios, thus enabling present actions and decisions. This type of capacity has a longitudinal scope, so that it is possible to speak of short-, medium-, and long-term anticipatory capacity. In addition, it should be remembered that not all uses of futures are equally relevant in terms of anticipation, since there are uses of contingent futures that have to do with a short- and medium-term reaction. There is another type of use of the future that has to do with the optimization of processes that are currently taking place in cities. Finally, there are exploratory uses of the future, which aim to significantly modify the trend of urban functioning and thus achieve future processes that are different from the current trend.

On the basis of the above, it would be possible to categorize the capacity of anticipation according to the temporal scope and the type of use of the future. In this line of thought, the peculiarities of the contingent use of the future force urban actors to make proposals with a short-term scope, conditions that limit their capacity to a low level. A medium capacity, on the other hand, has to do with the intention of optimizing the efforts made in the present in the future, which is why it tends to focus on the short to medium term. Finally, a high capacity for anticipation has to do with the use of the future, which seeks an exploratory future different from that proposed by the present trend, so it always pretends the long term.

The scope of anticipation offered by the *Todos Somos Juárez* program is medium to high. As mentioned above, the program seems to correspond to a process of breaking out of an undesirable present and the articulation of joint efforts (between government and citizens) to orient toward a better future; this implies that the focus was taken away from contingent anticipation and a style of exploratory anticipation was adopted, at least at the time. However, this intentionality was rather short-lived, and it was implemented in a rigid manner, so that once it was implemented, it stopped anticipating, causing a setback in the progress that had been made. In addition, the visualization process, although it tried to include as many aspects as possible, seems to have ignored the reaction of the criminal groups, so that when they changed their behavior, the project ceased to be effective and failed to achieve good results.

In terms of the anticipatory capacity of the Red Conéctate por tu Seguridad, this is identified as medium–low. Although the project does not intend to make a radical break with the present, there is a clear intention to modify the situation based on what citizens seek to improve in the future. It seems to recognize that the intended ideal is rather long-term, so no clear actions are established to achieve it, but it is established as a lighthouse that will allow to read the situation again in the future and to break this new present with another advocacy project. In other words, it corresponds to a long-term vision, but with contingent actions, in recognition of a limited advocacy capacity. In this sense, it favors an approach of replicable anticipation that encourages repeating the process as long as initiatives exist in the city. Unfortunately, there is no evidence of a third replication of the initiative after the projects that followed the implementation of the network (by MUCD).

Thus, in order to answer the guiding question of whether it is possible to speak of an urban anticipatory security system, this section examines the future literacy of different types of urban actors in the cases described above, as well as the application of this capacity to determine actions to be developed in the present. To this end, an analysis will be made based on three types of actors that can be found in a city: promoters of security, beneficiaries (or victims) of the urban condition, as well as detractors, those who benefit from a climate of insecurity. Each of these actors may have different capabilities to anticipate security in the same city, and these competencies are complicated by each other.

10.4.1 Security Promoters. Anticipation Capacity Medium–High

In order to interpret the anticipation capacity of the security promoters, it is necessary to recognize that the information on this task corresponds to strategic information of the States, so that even from reports and reports such as those presented above, they correspond only to a partial vision of the intelligence exercise carried out by the security promoters. However, the indicators of insecurity are recognized as an approximation of the efficiency of such processes.

In this sense, according to the information observed, it is possible to say that the capacity for anticipation, in terms of constructing future security scenarios, is also strong in terms of contingent future literacy and optimization. The security forces and governmental structures in charge of security build their scenarios from first-hand information, in addition to classified information, so they are the actors with the greatest amount of formal information on the phenomenon. Their anticipatory systems are strong in terms of medium- and long-term foresight.

Among its peculiarities, it is possible to mention that its prospective capacities formalize strategic information for the construction of possible scenarios and possible policies, anticipating situations of risk and vulnerability to crime and violence. By networking with other actors, it is able to extend its anticipatory capacity toward an exploratory logic. In addition, the way in which it determines unsafe and safe spaces depends mostly on the actions and strategies of criminals, so it is naturally forced to keep a large part of its anticipatory processes in reaction.

Its ability to use futures remains, at best, in the medium term. Its obligation to maintain the climate of peace, in addition to establishing itself as the agency that seeks to monopolize violence, keeps it in a permanent function of attention to expressions of insecurity. This keeps security planning permanently in the short term. This does not deny the possibility of establishing long-term scenarios by state actors, but given their short duration (based on political processes), it requires additional incentives to maintain this vision; a situation that in most cases does not happen.

On the other hand, the promoters of security are the actors who primarily collect information on acts of insecurity, as well as on measures and means to combat them. However, in a formal way, they can only relate to the beneficiaries of the programs, since the detractors act informally, in addition to establishing themselves as agents of competence in security matters. This leads to a partial identification of information.

This is particularly relevant in terms of anticipation. By referring to a construction process that goes beyond empirical information, operating from segmented information (intentionally) only complicates the anticipation process, since it is constantly confronted with anticipatory processes that seek to counteract the actions that seek specific scenarios.

10.4.2 Beneficiaries. Low Anticipatory Capacity

Security is not so much a result as a state that is experienced. In this sense, there are people who benefit from its expression in cities: the inhabitants. For them, their ability to anticipate security is linked to their ability to observe whether this condition is met in the environment in which they live. In this sense, the signals that individuals receive to interpret their future security situation are weak and diffuse.

The influence on the individual perception of urban security is constructed from incomplete information, as well as from unverifiable anecdotal information, so that the construction of the scenario of insecurity cannot be clearly formed. In addition, he is aware that moments of vulnerability (risk and threat) depend on the will of others rather than on foreseeable conditions, which prevents him from clearly determining the (un)safe space in the living environment. Thus, their capability to identify future scenarios is weak.

Their capacity to use and take advantage of their futures remains in the short term, because in the absence of specific incentives for this purpose, there are no incentives in the environment to operate them. On the one hand, the inability to formalize a clear scenario based on the signals he receives prevents him from proposing viable alternatives or modification mechanisms. In addition, the individual is limited in his ability to influence security; not only is he unable to use force when insecurity occurs, but he also does not have the resources or power to comprehensively address the structural problems that cause it.

For the beneficiary, it is impossible to give a sense of use to the information of the future, since one version is constantly pitted against the other. Understanding that security depends on a collective construction, the individual constantly observes how his perception of security is constantly confronted with information and scenarios proposed by other citizens, but also by promoters and detractors of security.

10.4.3 Detractors of Security—Anticipatory Capacity Medium–High

Although when we talk about the city and its capacities, we usually refer to the formal structures that make it up, the existence of individuals and groups that operate in a formal manner is also a reality. In terms of security, Ferrato (2014) recognizes the strategic capacity (both intelligence and structural) of criminals to influence conditions within a state. Within each city, this works in the same way. According to Arratia (2017), in spaces where insecurity phenomena operate, those who promote them and those who fight them find themselves in a confrontation of actions, strategies, and alternatives to win the loyalty of the inhabitants.

According to the National Security Strategy, the authorities in charge of security in México have enough problems to stop and dismantle criminal strategies, while the results seem to be an increase in insecurity. In addition, the document repeatedly acknowledges the infiltration and even complicity of public officials and government agencies. "As long as any authority struggles with political conflicts, crime advances in an organized manner" (Gobierno de México 2018, p. 10).

The above is not insignificant. Criminals operating in cities seem to act from processes very similar to those of formal structures, that is, they generate and make sense of strategic information. In addition to establishing actions and operations that reactively counter the actions of security promoters. In this sense, the prospective information with which they operate allows security detractors to exercise a strong capacity of anticipation.

With respect to the observed evidence, it is not possible to identify capabilities for future scenario building. However, the rapid response, as well as the crime and violence indicators obtained after the projects studied, point to a contingent and optimizing futures literacy. In addition, this type of actor has the capacity of cohesion over other actors. By linking with other actors, they are able to expand their anticipatory capacity toward an exploratory logic.

In the specific case of Ciudad Juarez, several analysts have considered the possibility that the efforts, resources, and actions developed by the Todos Somos Juarez program were not the reason for the change in the insecurity scheme, but that this could be explained by the activities of criminal groups (Arratia 2017; Sumano 2018). For its part, the evidence shows that in Azcapotzalco, despite the projects developed, the expressions of insecurity remained relatively stable.

The proactive nature of these actors means that their ability to use futures remains in the medium term, since security promoters, whatever their strategy, are forced to reduce it. This keeps security planning permanently in the short term, but also in the search for alternatives to expand its influence on the (in)security situation.

10.5 Anticipating the Future of Urban Security

As mentioned above, the reference to the future of security is usually interpreted as those actions of technification and prospective that the security agencies carry out in order to optimize their capacity for action. It is recognized that there are proposals in which artificial intelligence, the use of satellites and video surveillance systems represent the future of security, because in this way police agencies could improve their performance.

These approaches correspond to the traditional logic of understanding the anticipation of security in cities. However, the chapter also recognized that this approach does not consider the city as a complex system in terms of its actors, the objectives they pursue and the powers they have to achieve them. While there are government agencies that are strengthening their technological foresight capabilities, there are citizens who remain expectant of what is being done in terms of security. Moreover, there are actors who deliberately try to block the anticipatory capacities of security promoters by professionalizing a system of anticipating insecurity.

From the review of the evidence discussed in this chapter, the existence of anticipation by cities in terms of security is undeniable, but it is not as effective as the traditional approach shows. From the interaction of these three actors observed in the cases presented, it is possible to reflect on the anticipatory capacity of cities, which is more limited than we would like. As can be seen, the anticipatory capacity of the three different actors seeks to achieve different (in)security objectives, including limiting, hindering, or preventing the ability of the counterpart to anticipate its own actions.

In this sense, it is possible to confirm that security as a phenomenon is not only subject to very different structural social conditions, but also to the will of the actors involved in acts of insecurity. The formal responsibilities of state actors, as well as the interests of their opponents, allow these actors to exercise discretion in their response to (in)security. It is convenient for them not to provide information about their actions (e.g. high-impact security strategies or the strategies and actions of criminals). This favors a literacy about the future that varies between contingent and for optimization in general.

In general, the possibility of anticipating and understanding the future in terms of security by the urban system has been incapacitated by several factors, among which stand out: (a) an exponential growth of diffuse and sometimes contradictory signals, (b) a distancing of the motivations and capacities for intervention in security by the actors involved, and (c) a constant confrontation between promoters and detractors of security by the situation of crime and violence in the present and very short term.

The interplay of diffuse signals, as well as the inability to formalize a structure for constructing and using knowledge about the future in terms of security, has made it difficult not only for the authorities but also for the city's inhabitants to become literate in these terms. This makes it difficult to modify or even anticipate the security dynamics in which cities live; thus, the proposals and alternatives for intervention remain in the short term, in addition to favoring attention to the effects over the causes that generate them.

Although the cases studied present information that allows us to confirm the above, they also provide interesting evidence for reflection in terms of anticipation. One of the main peculiarities of *Todos Somos Juárez* was its implementation out of necessity. Faced with the recognition of an undesirable present for the city, it seeks to break with it by anticipating both desired and undesired possibilities.

The integration of different urban issues refers to the prediction of more than one outcome of the intervention. This is possible by examining the interaction of possible futures in order to select the most desirable one. The involvement of citizens and governmental actors at different levels also refers to a deliberation on the feasibility of the futures, which refers to an assessment of alternatives and an interpretation of the real scope that the futures would have from actions in the present. This allowed the practice of a more explorative reading of the future, although it was limited in its capacity to make use of this resource due to political restrictions.

On the other hand, the *Red Conéctate por tu Seguridad* project, although born out of the need to improve security conditions, does not represent such a marked break with the present, but rather a need to adapt the living conditions of the city. In this sense, the project was characterized by visualizing futures (short and medium term) and, based on a desirable future (long term), articulating actions to act in the present. Since we were not able to clearly visualize the results in the medium and long term, the tactic used was to sequentially rethink the logic of the project and adapt to the new present once it was achieved.

An important feature of the project has to do with its ability to establish the culture of legality as a guiding, but not central, element in the construction of intervention initiatives. That is to say, although the association promoting the project intended to train the participants in legality, it made it possible for the successive interventions and projects (now implemented by the beneficiaries of the workshops) to be adapted by each one of them. This opened up the possibility of seeing and proposing futures based on the observations of each of the participants.

Although the Municipality of Azcapotzalco was one of the promoters of the project in the city, there is little evidence of the use and application of the information generated in terms of futures. In this sense, although an alliance between the anticipatory capacities of both the promoters and the beneficiaries of security is identified, it is not possible to verify that they have worked together and have had results.

Moreover, speaking of an urban system of anticipation forces us to think not only about state interventions, but also about the interaction that these interventions represent in the city as a whole. In this sense, the evidence allows us to observe a very similar capacity for anticipation between the security detractors and the security promoters. Although security benefits were obtained during the implementation of the program, the quick reaction and adaptation of the criminal groups allowed a setback in this situation. Their adaptation of strategy is evidence of their ability to explore new futures based on contingency, which overcame the initial strategy of the security promoters.

10.6 Final Thoughts

The available information on the complex relationship between crime increase and decrease in both cities only reflects the high adaptive capacity of both security promoters and detractors. In both cases, after the change in the (in)security situation, a change in the opponents' strategy was observed, leading to a return to the initial situation. Beneficiaries, on the other hand, remain with a low anticipatory capacity, as they generally continue to search for alternatives in the short term and begin to operate in a contingent manner.

According to the evidence presented, this is related not only to their anticipation capacities, but also to their formal and informal functional structures. The former operate on the basis of rigid mandates and during periods that interrupt medium- and long-term processes. On the other hand, the second type prefers to operate on a discretionary basis and therefore has an interest in maintaining a dispersed and unclear system of signals about the future.

It can be said that the cities studied have been able to structure and give meaning to the anticipatory systems in a functional way for their operation, but keep them in a medium capacity. However, the peculiarities found give important information about the possibility of expanding and operating urban anticipatory systems in terms of security. The positive results in both cases allow us to reflect on the virtues and contributions of anticipation to urban security.

Recognizing the complexity of actors, objectives, and mechanisms that exist within cities to address security strategies is relevant to improving urban anticipation. An urban anticipation system cannot consider only one of its elements, but must include them or, in the worst case, consider the existence of the rest. The issue of urban security is particularly challenging, since the actors not only have different objectives, but there are also marked oppositions, which implies the will to obstruct and deny the possibilities of the other to achieve the desired future.

In this sense, it is undeniable that the technological tools that have been introduced for security agencies are bringing us closer to this goal. Information management mechanisms and new institutional structures allow us to imagine scenarios in which anticipation can be strengthened throughout the city. Security agencies must continue to strengthen their prospective and adaptive capacities in order to visualize scenarios, so that today it is possible to develop actions that are neither reactive nor traditional prevention, but rather transform urban conditions.

An indispensable consideration for a dialogue on the urban future of security has to do with observing the models and tools of prospective with a lens that recognizes the distortion generated by the existence of a strategy that pretends to anticipate in turn, but with the opposite ends in mind. That is, to recognize the game of anticipation that both security agencies and those who seek insecurity seek. In this aspect, it seems essential to involve the inhabitant actor in order to provide additional information on these future scenarios.

References

Aguirre J (2016) La tecnología de información y comunicación en prevención del delito. URVIO 18:90–103

Alcaldía Azcapotzalco & MUCD (2021) Reporte final de actividades de intervención y resultados 2019–2021. Retrieved 15 Nov 2023, from https://www.mucd.org.mx/wp-content/uploads/2021/10/Reporte-RCxTS-Azcapotzalco-2019-2021.pdf

Alvarado N (2018) Tecnología contra el crimen: Entusiasmo con cautela y criterio. Sin miedos. Seguridad Ciudadana. https://blogs.iadb.org/seguridad-ciudadana/es/tecnologia-contra-el-crimen-entusiasmo-con-criterio/

Aquilué I, Ruiz J (2021) Ciudad, complejidad y cambio: fundamentos para el análisis de la incertidumbre en sistemas urbanos. Revista INVI 36(101)

Arratia E (2017) Todos Somos Juárez. Competititon in state-making y la guerra contra el narcotráfico (2006–2012). Rev Esp Cienc Polít 43:83–111. https://doi.org/10.21308/recp.43.04
Arriagada I (2001). Seguridad ciudadana y violencia en América Latina. In: LASA 2001 XXIII international congress session, Washington
Arriagada I, Godoy L (1999) Seguridad Ciudadana y violencia en América Latina: diagnóstico y políticas en los años noventa. CEPAL. Serie de Políticas Sociales 32, LC/L. 1179-P
Bauman Z, Donskis L (2015) Ceguera moral. La pérdida de sensibilidad en la modernidad líquida. Paidós
Boer A, Wiekens C, Damhof L (2018) How futures literate are you? Exploratory research on how to operationalize and measure futures literacy. In: 6th international conference on Future-Oriented Technology Analysis (FTA)—future in the making Brussels
Brenner N (2013) Tesis sobre la urbanización planetaria. Nueva Soc 243
Carreón J, García C (2014) Teorías de la seguridad pública y percepción del delito. Margen 71
Csernatoni R, Oliveira B (2023) Disruptive technologies for security and defence: temporality, performativity and imagination. Geopolitics. https://doi.org/10.1080/14650045.2023.2224235
Ferrato A (2014) Crimen organizado e inteligencia estratégica: desafíos para el siglo XXI. URVIO 15:55–70. https://doi.org/10.17141/urvio.15.2014.1587
Foucault M (2006) Seguridad, territorio, población: Curso en el Collège de France: 1977:1978. Fondo de Cultura Económica. Buenos Aires, Argentina
Gobierno de México (2018) Estrategia Nacional de Seguridad Pública. Retrieved 15 Nov 2023, https://www.gob.mx/cms/uploads/attachment/file/434517/Estrategia_Seguridad-ilovepdf-compressed-ilovepdf-compressed-ilovepdf-compressed__1_.pdf
Gobierno de México (2021) Factores de Inseguridad. Observatorio de reclutamiento. Retrieved 15 Nov 2023, https://www.gob.mx/sspc/observatorioreclutamiento/documentos/factores-de-inseguridad
González-Arellano S (2022) The future scenarios of cities: an analysis of their institutional construction abstract. Int J Urban Reg Res 46(5):865–873. https://doi.org/10.1111/1468-2427.13103
Habegger B (2022) Securing the future: the use of strategic foresight in the security sector. Geneva Centre for security policy. Strateg Secur Anal 23. https://doi.org/10.21256/zhaw-28391
Herrera-Lasso L (2013) Factores que propician la violencia y la inseguridad: apuntes para una estrategia integral de seguridad pública en México. CEI-El Colegio de México, México. Retrieved 15 Nov 2023, https://cei.colmex.mx/archivos/213/Luis_Herrera-Lasso__Factores_que_propician_la_violencia_y_la_inseguridad.pdf
Krauze K (1998) Theorizing security, state formation and the 'Third World' in the post-Cold War world. Rev Int Stud 24(01):125–136. https://doi.org/10.1017/S0260210598001259
Lechner M (2016) Tecnologías aplicadas a la seguridad ciudadana: desafíos para la justicia transicional ante nuevos mecanismos de control social. Divulgatio 1(1). https://doi.org/10.48160/25913530di01.9
Montero J (2013) El concepto de seguridad en el nuevo paradigma de la normatividad mexicana. Reg Soc XXV(58). https://doi.org/10.22198/rys.2013.58.a128
MUCD (2019a) Cultura de la legalidad Centro Histórico. Polígonos Santa María de la Redonda y la Antigua Merced. Retrieved 15 Nov 2023, https://www.mucd.org.mx/wp-content/uploads/2017/10/ESTADO-DE-ME%CC%81XICO-2015-TUI.pdf
MUCD (2019b) Azcapotzalco. Plataforma de Participación ciudadana 2019–2021. Retrieved 15 Nov 2023, https://www.mucd.org.mx/azcapotzalco/
Muggah R, Szabó de Carvalho I, Alvarado N, Marmolejo L, Wang R (2016) Haciendo de las ciudades lugares más seguros: Innovaciones sobre seguridad ciudadana en América Latina. Artículo estratégico 20 de junio 2016. Ed. Instituto Igarapé, Banco Interamericano de Desarrollo y Consejo sobre Fragilidad, Conflicto y Violencia de la Agencia Global del Foro Económico Mundial
Paniagua A, Camargo-González I (2017) La implementación de las políticas públicas para la prevención de la violencia en el municipio de Ciudad Juárez, Chihuahua: el involucramiento de la

sociedad civil de los liderazgos comunitarios. Acta Universitaria 27(4):92–108. https://doi.org/10.15174/au.2017.1457

PNUD (1994) Informe sobre desarrollo humano 1994. Fondo de Cultura Económica, México

Poli R (2019) Introducing anticipation. In: Handbook of anticipation theoretical an applied aspects of the use of future in decision making. Springer. https://doi.org/10.1007/978-3-319-31737-3_1-1

Rosas T (2012) Replican Todos Somos Juárez en Acapulco. El Economista. 31 de mayo de 2012

Sumano J (2018) La violencia en Ciudad Juárez: a seis años de la estrategia Todos Somos Juárez. Nexos—Paz Seguridad, Julio 2018. Retrieved 15 Nov 2023, https://seguridad.nexos.com.mx/la-violencia-en-ciudad-juarez-a-seis-anos-de-la-estrategia-todos-somos-juarez/

Vergel-Clavijo G, Guerrero-Bayona A (2023) Ciudad inteligente: mejoramiento de la seguridad a través del uso de nuevas tecnologías. Rev Ingenio 20(1). https://doi.org/10.22463/2011642X.3510

Vistazo al futuro (2021) Se activa "Red Conéctate por tu seguridad" Azcapotzalco. 5 de marzo de 2021 Retrieved 15 Nov 2023, https://vistazoalfuturo.com/se-activa-red-conectate-por-tu-seguridad-en-azcapotzalco/2785/

Williams M, Axon L, Nurse J, Creese S (2016) Future scenarios and challenges for security and privacy. In: IEEE 2nd international forum on research and technologies for society and industry leveraging a better tomorrow (RTSI), Bologna, Italy, pp 1–6

Chapter 11
Anticipating Zero Plastics Futures in Urban Contexts in Mexico

Bruno Gandlgruber and Marco Jaso

Abstract The production and consumption of plastics is closely related to the prosperity, but also the calamities of the modern socioeconomic system in terms of the generation of excessive waste. And cities have been considered the primary origin of plastic contamination of all land and marine ecosystems. In response, different stakeholders at a global and local level have used diverse images and narratives to create scenarios of the future to illustrate the long-term consequences of plastic pollution, but also strategies of mitigation and adaptation. In this chapter we explore some of the main contributions, following the conceptual distinction between foresight, forecast, and anticipation in contemporary futures studies. We are particularly interested in narratives related to long-term efforts to radically reduce plastic pollution in urban contexts, also called zero plastics futures, and their influence on international regulatory efforts to cope with this challenge. In the last section we develop some elements for a fundamental approach to conceive the urban plastic economy as an anticipatory system to achieve a deeper understanding on the characteristics of the role of urban anticipatory communities. We use a brief overview of important advances in narratives of regulation for the mitigation of plastics on a national and city level in Mexico, with particular emphasis on the relation to zero plastics.

Keywords Anticipatory systems · Anticipatory communities · Future scenarios · Plastics economy · Zero plastics policies

B. Gandlgruber (✉) · M. Jaso
Department of Institutional Studies, Universidad Autónoma Metropolitana-Cuajimalpa, Mexico City, Mexico
e-mail: bgandl@cua.uam.mx

M. Jaso
e-mail: mjaso@cua.uam.mx

S. Gonzalez-Arellano and B. Gandlgruber (eds.), *Cities as Anticipatory Systems*, The Urban Book Series, https://doi.org/10.1007/978-3-031-78162-9_11

11.1 Introduction

Plastics have many important properties for a wide variety of applications. Still, the socially and economically established practices of their use, mainly because of the intensive processing of fossil fuels and the inadequate forms of plastic waste disposal, have generated a wide variety of health and environmental risks, which has led to a growing consensus on the importance to reduce its production and consumption and to transform the sociotechnical system that it sustains toward a configuration that mitigates these risks. The transformation of this system toward a significant reduction of the negative impacts is a complex problem on a global scale that involves many different actors in different social spheres.

This chapter analyzes different aspects of this problem. It develops an initial analytical proposal based on the urban plastic economy as an anticipatory system that creates images, narratives, and models of a future with a strictly reduced and modified form of plastic use in urban contexts. We critically analyze the use of the future in documented proposals of this kind of transformation process. Capacities of anticipation refer not only to strategic planning and foresight by core decision-makers like multinational companies and environmental agencies in different levels of government but also to the required societal potential of imagining and applying changes on the social institutions or rule systems as guided by future narratives. As the primary origin of plastic contamination of (land and sea) ecosystems, cities play a central role in these transformation processes.

The chapter includes four main sections. We begin in the first section with a general explanation of the expansion of the plastics economy as a complex social system of production and consumption in recent years and the growing risks to health and the environment because of the often uncontrolled diffusion of plastic debris into the environment. In the second section, we analyze the particular uses of images, narratives, and models of the future related to the plastics economy, following the distinction of the three levels of future studies: foresight, forecast, and anticipation. At the end of the second section, we develop a fundamental approach to conceive the urban plastic economy as an anticipatory system. In the third section, we give a brief overview of important advances in narratives of regulation for the mitigation of plastics on a global and national level in Mexico, with particular emphasis on the proposal of zero waste or zero plastics as a scenario where plastics are completely irradicated at some moment in the future. The fourth section looks at this process of building rules as the institutionalization of futures in regulating plastic use at the level of Mexico City. The information we gathered allows us to characterize different aspects of the urban plastics economy in Mexico as an anticipatory system, like elements of its structure and specific anticipatory capacities of visualization and transformation.

11.2 Plastics, Plastics Pollution, and the Plastics Economy

Plastics as non-natural polymeric systems[1] like polyethylene, polyester, or polypropylene are considered one of the major human inventions in modern times, and they are ubiquitous in contemporary production and consumption systems for purposes like storage, transportation, packaging, or insulation. Their flexibility, adaptability, and resistance have made it possible to substitute wood, metal, or glass at lower production costs (Andrady and Neal 2009). These societal benefits of plastics explain their high popularity, and new properties and applications in areas like energy conservation, robotics, or transplantation medicine are continuously added.

However, the exponential growth of plastic applications in modern times[2] has severe downsides. The unconstrained disposal of plastics has created growing levels of environmental stress and has become a global threat to all living and non-living systems (Iroegbu et al. 2021, p. 19344). We are talking about different forms of **plastic pollution** (macro-, micro-, and nano-plastic debris) as "the intrusion or invasion by plastic materials (i.e., polymeric systems), either through direct introduction or degradation processes of environments (to which they are not native) to negatively or undesirably impact such environments" (ibid.). Plastic pollution is an anthropogenic form of pollution through the introduction of plastic materials in different spheres of the environment that are not "natural" to them. They cause (reversible or irreversible) modifications, disturbances, and disruptions of ecosystems and balances between ecosystems (ibid.).

The hazardous nature of plastic pollution is a consequence of the scale of production, its biological impact at low levels of concentration, the diffusion and degradation of plastic debris in air, water, and solid textures, its long-term endurance, and the diverse consequences for human health and the stability of the invaded ecosystems. Due to the complex nature of these processes and the growing visibility of diverse repercussions, the number of scientific publications related to "plastic pollution" has grown significantly in recent years to reach almost 460 thousand in the last decade (ibid.).

---

[1] There are also natural polymers like cellulose, carbohydrates, proteins or rubber, and human DNA or skin (Iroegbu et al. 2021, p. 19345).

[2] Garbage, in particular household garbage, is quite a recent phenomenon. Until the nineteenth century, leftovers and broken objects had usually specific uses or were repaired: "Leftover food was boiled to make soup or fed to animals. Things that lost their use for adults became playthings for children, broken objects were repaired, and durable goods and items, like furniture, clocks, and watches, were handed down from generation to generation. In agriculture and animal husbandry, manure was used as organic fertilizer, straw was reused in construction, rags found a new purpose in papermaking, bones were valued in producing soap, and ashes were used to control pests and fertilize soils. Even dog excrement was a sought-after commodity, as it served a purpose in tanning leather. For thousands of years, economies were circular economies: they produced very little waste (…). This began to change with urbanization and rising populations and, more importantly, with the advance of assembly-line manufacturing that made cheap, mass-produced items available worldwide. Consumerism created a throw-away society, particularly in the United States, that produced an excess of easy-to-use, disposable items" (Mauch 2016, p. 5).

We can distinguish three categories of plastics in the global "plastics ecosystem": plastics in use, managed plastic waste, and mismanaged plastic waste (Lebreton and Andrady 2019). For example, 11% of the global mismanaged plastic waste generated in 2016 (19–23 million metric tons) found its way into different aquatic environments, with growingly disastrous consequences (Jambeck 2015). Continuous growth of these amounts to approximately 53 million metric tons per year in 2030 is projected, even in scenarios of effective regulation. Without the application of restrictive policies, it could even reach 80 million metric tons by 2030 (Borrelle et al. 2020, p. 1516), with an intense concentration in specific regions in Asia and Africa with the highest levels of demographic and urban growth (Lebreton and Andrady 2019).[3] One of the main problems is the insufficient solid waste infrastructure, even in industrialized countries with high levels of waste generation,[4] and the massive exportation of plastic waste to countries with low-quality standards and recycling capacities, which often just dump or burn plastic garbage indiscriminately (Law et al. 2020). All these factors contribute to the fast growth of plastic intrusion into the environment.

The main **sources of plastic pollution** are different types of plastic waste and debris, mainly contained in textiles and fibers, cosmetics, paints and coatings, and many other basic consumer products (plastic bags, toothbrushes, shoes, electric cables, writing pens, mattresses, wipes, computer casings, etc.). One of the several particularly invasive forms of microplastic debris is tire abrasion ("wear and tear") from all kinds of vehicles. On average, the global per capita emission of microplastics through tire abrasion is estimated to have reached more than 800 g yearly (Kole et al. 2017). An estimated 5–10% reaches the oceans from the microplastic debris generated through tire abrasion, and 3–7% is diffused in the air (ibid.). Tire abrasion is also a form of contamination that has been difficult to address, partly due to the lack of comprehensive studies and partly due to the intense lobbying of the powerful tire and car industries against regulation (ibid.). This is also illustrated through a study by the Swedish government from 2020 that showed that more than 50% of Sweden's microplastic pollution is a consequence of tire abrasion. However, a broad counter-initiative by these industries has blocked stricter measures proposed by the European Union (European Commission 2021, Iroegbu et al. 2021, p. 19348).

The specific consequences of plastic pollution are often still not known in detail because of a lack of long-term impact studies, despite the growing research interest, but what we already know so far indicates a wide range of calamities from inhalation of microplastics, intrusions from apparel and home textiles, as well as microplastics contained in drinking water. There are mainly two types of problems: **health risks and environmental risks**. In general, 20% of global deaths are estimated to be related to pollution-caused health complications (Prüss-Ustün et al. 2016). In particular, a rather long list of health problems, like thyroid dysfunction, obesity, diabetes, and reproductive impairment, have been related to different forms of plastic pollution

[3] More details on future scenarios in the following sections.

[4] The US as the country with the highest amount of plastic garbage generated 42 million metric tons in 2016 (Law et al. 2020).

(Iroegbu et al. 2021, p. 19346). Some authors have recently pointed to the negative impact of nanoplastics on microorganisms in the digestive system, which are, among other things, responsible for establishing chemical balances in the endocrine, immune, and nervous systems (Teles et al. 2020). Additionally, there are indications that micro- and nanoplastics can cause specific forms of genotoxicity (ibid.). The main environmental risks related to plastic pollution refer to the acceleration of climate change through its fossil origin and plastic contamination of aquatic environments, the acidification of seawater through dichlorination of PVC, and other (sometimes hard to anticipate) effects of plastic debris in ecosystems (Iroegbu et al. 2021, p. 19347).

To provide and distribute the vigorously growing quantity and variety of plastics, an extensive and diversified **plastics economy** has emerged as a complex sociotechnical system that organizes interconnected value chains of innovation, production, consumption, and recycling of plastics, including the formation of institutional frameworks of diverse social rules (habits, laws, regulation, and others).

Various **policies** and tools to reduce plastic pollution have been developed and partially applied in recent years. The common objective of all these efforts has been to control and mitigate the generation and diffusion of plastic debris into the environment. Some policies focus mainly on the production side, promoting reducing the use of polymers and their substitution with green(er) materials like bioplastics and biodegradability (UNEP 2018; WRI Mexico 2020; OECD 2022a). A related approach aims to strengthen extended producer responsibility with voluntary contributions and different forms of circular economies (Lindhqvist 2000; Ellen McArthur Foundation 2021; GAIA 2022). The third group of mitigation measures emphasizes consumer knowledge and responsibility through promoting a green lifestyle, enhanced education efforts, and a wider diffusion of scientific studies on the matter (Iroegbu et al. 2021, pp. 19350–19351). Most of these efforts to control, mitigate, and regulate plastic pollution are related to urban contexts of production and consumption. A comprehensive example of an integrative policy is the EU program of phasing out the main single-use plastic products,[5] with specific targets for separation, recycling, and bans[6] (EU 2021).

Several obstacles can be associated with the application of existing policies and tools. First, the narrow focus of measures on more visible topics like plastic bags and other single-use plastic items has obscured the public view on other equally or even more critical problems. As mentioned, the second problem has been the trivialization through apparently diverging scientific positions regarding the causes of environmental affectations and the associated levels of risk. Finally, the asymmetry

[5] "Target the top single-use plastic products found on our beaches: bottles (and caps); cups (and lids); cutlery, plates, straws, and stirrers; food containers; lightweight plastic bags; packets and wrappers; cigarette butts; tampons, applicators, sanitary pads, and wet wipes; and cotton buds, balloons and balloon sticks. Also, aim for oxo-degradable plastic products and fishing gear containing plastic" (EU 2021).

[6] "Specific targets include a 77% separate collection target for plastic bottles by 2025—increasing to 90% by 2029, incorporating 25% of recycled plastic in PET beverage bottles from 2025, and 30% in all plastic beverage bottles from 2030" (EU 2021).

of power of the different stakeholders involved, with particular interests of multinational companies and their strong influence on the research agenda and also on the political decision-making process, has weakened the responses (Iroegbu et al. 2021, pp. 19347–19348; Dumbili and Henderson 2020).

11.3 Conceiving Futures of the Plastics Economy

Due to the widespread and growing consequences of plastic pollution and the associated (mostly hidden or external) costs of today's approximately USD 1 trillion related to the loss of ocean productivity alone and USD 2 trillion of the total cost (OECD 2023, p. 24), the topic has been included in the catalog of UNESCO's social development goals (SDG) for 2030, generally expressed in SDG 14: "conserve and sustainably use the oceans, seas and marine resources for sustainable development" and the specific objective SDG 14.1, that demands "the prevention of plastic pollution through sustainable development" (UN 2015, p. 23). The formulation of these specific objectives that describe pathways of change toward lower pollution levels is closely related to the construction of future scenarios of adapted production and consumption systems based on different assumptions and methods of futures thinking involved.

Plastic pollution is a long-term global issue[7] that can only be tackled with a transformation at all levels of the plastics economy. The impact of plastic pollution is already widely perceived and documented today. However, the central health and environmental risks will unfold due to the projected growth of plastics production and the prevalence of low capacities to reduce uncontrolled environmental diffusion (mismanaged plastic waste). The development of the plastics ecosystem in general, the evaluation of the associated risks, and the capacity to create and apply measures for a sustained transition process are closely related to open future developments and to the diverse **representations of possible futures**, including the disposition to enhance the collective capacity to push the system toward more positive, sustainable futures. By creating and using different images of its future, the plastics economy or the diverse coalitions and communities, particularly in urban contexts, can be considered anticipatory systems. When we analyze the characteristics of an anticipatory system, following the general conceptual framework that guides this book (see Chap. 1 by Gonzalez), we are interested in these images and in the social processes of creating and diffusing them, as well as the consequences for individual and collective decision-making and the related construction of narratives about alternative futures and possibilities to build preferable futures. In this process, formal institutions and regulations play a vital role in conveying those narratives to stakeholders by setting goals and points of arrival.

That is why we are recurring to different approaches of contemporary futures studies that recognize the substantial prevalence of uncertainty in all efforts to deal

[7] Even if we would stop the use of plastics immediately, the consequences of past pollution persist.

with the future and that combine a whole variety of methods as applied in the three basic levels of analysis: forecast, foresight and anticipation (Poli 2019). There is no precise temporal or conceptual separation or temporal succession of these three approaches in the canon of futures studies since there are different overlaps and combinations of diverse methods.

Forecast, as the first and basic level of futures studies, is characterized by efforts to predict the specific behavior of a phenomenon or a social system in the future, based mainly on quantitative analysis from historical data related to selected variables. The extrapolation of behaviors from the past to the future expresses (at least implicitly) strong assumptions about the continuity of the corresponding social systems with a somewhat limited capacity to consider systemic uncertainty, the possibility of ruptures in the future, and the mental and social capacities of pushing alternative agendas. Well-known examples of the forecast type of future studies include the oil peak studies about the prediction of the moment of maximum fossil consumption or business-as-usual climate scenarios that predict a specific rise of the global and regional temperatures as a consequence of the linear growth of greenhouse gas emissions. Forecast models represent the early stages of futures studies in the 1960s and 1970s and are still extremely popular in many scientific studies and policy proposals. They are, of course, important inputs for contemporary futures studies and decision-making because they allow us to visualize the consequences of linear dynamics in stable systems. Nevertheless, additional tools are needed to take into account possible ruptures and changes in the mental models of the agents that compose the social systems, and these are crucial elements to understand and guide all kinds of transition and transformation processes.

Projections on the level of foresight exercises are widespread in many studies on the plastics economy. The statistical data that estimates future developments based on static economic models and linear anticipatory assumptions generally predict strong growth of plastics production for the next decades. Following OECD data (OECD 2022b), global production from 2020 will approximately double until 2040 at 800 million tons per year and reach 1.2 billion tons in 2060. Foresight calculations for mismanaged plastic waste show similar tendencies, reaching 200 million tons in 2060 (Lebreton and Andrady 2019). This linear overall tendency in plastic debris results from a moderate reduction in the Global North combined with a relative growth in the Global South, with a high concentration in African and Asian countries, following the estimated population growth rates and urbanization in these regions. One lesson from these projections is to promote large-scale investment in waste management in these countries (Kaza et al. 2018). These measures could have a determining impact in vulnerable countries where the populations are strongly affected by health and environmental risks (ibid.) and also allow for reducing CO_2-equivalent emissions. Projections are an important form of creating images about the future because they allow us to understand what might happen if nothing changes. For that same reason, the associated anticipatory value is limited because there is very little space to conceive alternations to linear extrapolation. Scenarios outside these limited trajectories, which can be important to understand risks and opportunities, are often blind spots in the futures map.

Strategic foresight as the second level of future studies has incorporated several qualitative methods, like exploratory scenario analysis or weak signals, among many other innovative methods (Schwartz 2023), that allow for a much broader perspective in long-term analysis, with many implications for a more robust understanding of possible pathways in the future development of all kinds of social systems. More awareness of less probable existential risks can strengthen individual and social resilience; imagining diverse nonlinear opportunities helps conceive preferable futures and liberate significant potentials for their realization.

Foresight exercises are also quite common in conceiving future images of the plastics economy, generating alternative future scenarios with mixed methods. For example, comparing possible scenarios for plastic emissions into aquatic systems allows us to anticipate the consequences of the transition process of the plastics economy and the guiding rule systems that must be developed (Borrelle et al. 2020, see Fig. 11.1). These approaches imagine different pathways of the plastics economy regarding production levels, types of plastics used, and recycling capacities. The foresight exercises, like the ones we introduced in the chapter before, are closely related to the "business-as-usual" (continuity) type of scenarios. Borrelle et al. (2020) additionally use intervals of confidence that reflect the critical effects of dispersions that are part of a more profound scenario analysis, where uncertainty is taken for granted and specific anticipatory assumptions are included. The baseline scenario adapted from the World Bank leads to growth dynamics of aquatic plastic pollution, with approximately 50 and 90 million tons in 2030. A second scenario (called "ambitious") estimates a reduction of plastic debris that reaches aquatic systems to 20–50 million tons in 2030 if all countries comply with a series of global and national commitments, like those included in the G7 Plastics Charter. This would mean reaching national targets of between 30 and 90% of managed waste, depending on the classification in the Plastics Charter. In terms of futures studies, the "ambitious" scenario implies that a systematic and effective "formal" regulation of plastic use on a global level would have to be put in place, but also that images and narratives of the underlying future images are being shared and accepted. This point is even more compelling with the third scenario, the so-called "target" scenario, which is based on a *backcasting* logic of formulating pathways and specific strategies that allow reaching a pre-defined target of 8 million tons of pollution per year as an amount that is necessary for the long-term recovery of marine ecosystems. The future images and anticipatory assumption of this type of scenario not only follow a different logic of construction but also question the direction of the long-term development of the plastics economy and the economic model in general, introducing goals of a general reduction of 85% of plastics production and high levels of recycling, that would require a reorganization of many related economic subsystems toward effective sustainability and alternative forms of growth.

When we analyze the plastics economy with strategic foresight tools, we can recognize alternative visions of transformation and outline particular strategies and tools (regulation, bans) or specific roles for different actors in futures scenarios delineating alternative scenarios (Greenpeace 2019a, b). Other proposals have created narratives of the future that connect technological innovation related to smart cities

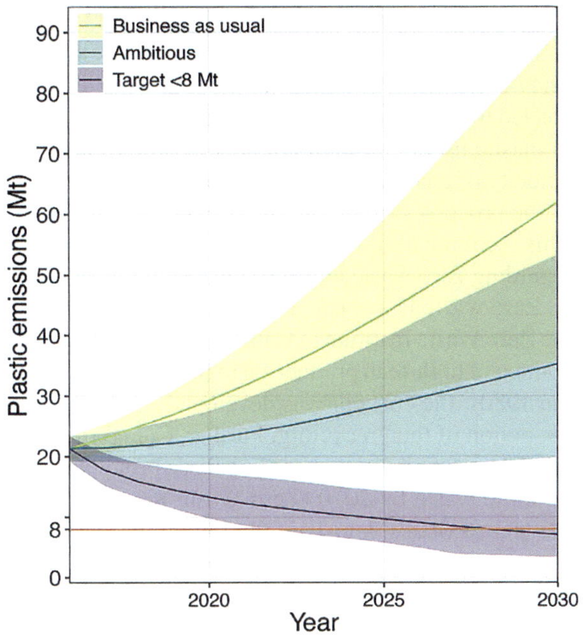

Fig. 11.1 Strategic foresight and future scenarios for the plastics economy. *Source* Borrelle (2020), p. 1516. Reprinted with permission from AAAS

with new forms of urban waste management (Circular Economy 4.0) (Mavropoulos and Nilsen 2020).

From an alternative epistemic perspective, the **theory of anticipation** (Poli 2019), as the centerpiece of the third level of futures studies, systematically analyzes different uses of the future in anticipatory systems that are part of any existing social system and explores the conditions for generating alternative futures based on the collective intelligence of empowered and more "future literate" actors (Miller 2018). The most considerable differences between forecast, foresight, and anticipation lie in the human conceptualization of uncertainty and complexity, the importance of an open and emergent construction of desirable futures, and the participatory and interactive governance of visualization and transformation processes. Anticipation theory revises mental models, imaginaries, and narratives of the future with novel analysis and practice tools. The first step in the participatory analysis of anticipatory systems is characterizing their images and the main processes of their creation, including mental models with all their biases and prejudices. Frequently, participating actors recognize the prevalence of low anticipatory capacities, linear anticipatory assumptions, and narrow pathways for evolving transition processes. In the second step, existing knowledge repositories and institutional rule systems are modified and reframed to explore trajectories that lead to scenarios with more positive perceptions. One of the main differences with strategic foresight is that there is no preestablished "best" alternative.

11.4 Zero Waste and Zero Plastics Futures Initiatives

The development of images of the future related to the reduction of plastic pollution has led to dynamics of imagining futures without plastic use, a significant reduction in plastic use, or without plastic pollution. The differences in the scenarios and the velocity of change depend on other elements of economic and social change related to the processes of socioecological transition (Fischer-Kowalski and Haberl 2007). This section will illustrate some of these narratives and their differences, particularly regarding zero waste and zero plastics movements.

Zero waste is a concept coined in 1970 by Paul Palmer, the founding director of the Zero Waste Institute in California, highlighting the economic value of waste in the context of high-tech production in Silicon Valley (Mauch 2016, p. 6). Later on, during the 1980s, the concept was developed further by a group of European experts with the notion of total recycling. Further, Canberra in Australia was the first city that, in 1995, began to legislate based on these principles and set itself the goal of generating no more waste by 2010 (Zero Waste Revolution 2024). Also, in San Francisco, in the United States, legislation based on the same principles was introduced in 2002 and was crucial for reducing waste by more than half in a decade (EPA 2024). The zero waste movement created a global platform called the Zero Waste International Alliance (ZWIA) to share experiences in different cities worldwide and establish common standards to guide the development of zero waste projects worldwide (Zero Waste International Alliance (ZWIA) 2019). It evokes a different scenario and has begun integrating a broad international movement to redirect public policy initiatives toward a comprehensive and preventative approach inherent in the circular economy paradigm (ZWIA 2017; Hannon and Zaman 2018; Brandao et al. 2020; Baskar et al. 2022).

In recent years, zero waste has become a catchword for waste reduction in companies, cities, and countries worldwide. However, the specific applications refer to rather different problems and objectives that sometimes differ from a systematic strategy of avoiding plastic debris. In all these cases, zero waste refers to an image of a future free of (unnecessary) waste and particularly free of plastic waste. Zero waste is the central part of a future narrative to build social systems capable of organizing production and consumption without the negative impact of plastic pollution. Many different actors have contributed to creating and diffusing this narrative and developing and implementing different strategies for plastic waste reduction through the transformation of the plastics economy on different levels, particularly in urban contexts, with the creation of communities like grassroots foundations, zero waste communities or citizens' movements in different parts of the world (Kaza et al. 2018; UNEP 2018, 2020).

The basic narrative of the zero waste concept promoted by a diverse group of urban communities and movements is related to changing production and consumption practices that allow the reduction of waste with different forms of feedback into the economic cycle, preventing it from being released into the environment. Zero waste means in the first place designing and managing products and processes to systematically avoid and eliminate the volume and toxicity of waste and materials, conserve and recover all resources, and not burn or bury them, and is, in that sense, closely related to the idea of a circular economy. A systematic "zero waste" approach implies not just building technical solutions for recycling plastic waste but also redesigning products and production processes, allowing the reuse of all materials along the whole value chain. In that sense, it is related to images and narratives of the future economy associated with transforming large parts of the production and consumption system. This means creating a different mental model for producers and consumers or a "zero waste philosophy" (Mauch 2016, pp. 6–7). "Zero-waste is a goal that is ethical, economical, efficient and visionary, to guide people in changing their lifestyles and practices to emulate sustainable natural cycles, where all discarded materials are designed to become resources for others to use" (Zero Waste International Alliance 2024). These initiatives are closely connected to other processes of institution building at different levels.

At a global level, the United Nations Environment Organization (UNEP) and the World Economic Forum have used future risk assessment methods to mitigate the global problem of the distribution of inadequately managed urban plastic waste through bodies of water and air currents. For example, a 2017 UN resolution signed by nearly 200 nations stands out, emphasizing the importance of tackling the future growth of ocean plastic pollution (UNEP 2022). When China suspended the massive imports of plastic garbage from major industrialized countries in 2018, efforts to promote circular plastic management accelerated. The Ocean Plastic charter mission, signed by the G7, stands out. The European Commission published the "European Strategy for Plastics in the Circular Economy," and the OECD held the "Global Forum on Plastics in the Circular Economy." One of the most relevant milestones within the framework of the 5th UNEP Assembly was the resolution "End Plastic Pollution: Towards an internationally legally binding instrument." The Intergovernmental Negotiating Committee is expected to arrive at a legally binding treaty by the end of 2024, with a comprehensive circular economy approach, encouraging the participation of all stakeholders and promoting cooperation at all levels. In the meantime, stakeholders are engaged in periodic discussions and negotiations on the final draft intended to be voted on in the 6th UNEP Assembly in early 2025 (UNEP 2022).

Public pressure has pushed federal and local authorities related to environmental, industrial, and waste management policy to pay more attention to plastic pollution as an environmental policy priority. Policymakers have experienced the progressive emergence in their agendas of the measures to stop plastic contamination (UrbanShift 2023a, b). At the same time, relevant sectors of the population, with greater capacity for agency, showed their concern in social networks and news media regarding the gigantic islands of plastic parked in the ocean gyres, the images of marine fauna suffocated by plastic, the pieces of plastic found in the stomachs of fish and birds, as well as the videos of severe affectations of marine fauna (2015) (Barkela et al. 2021; Teh et al. 2022; Pop et al. 2023).

These initiatives have become more present, but they have had a rather limited impact, particularly in countries in the Global South. Among different challenges, we note that international policies and social norms are pending to be introduced and applied more systematically. Formal institution building is advancing but is not enough. We argue here that the existing anticipatory systems are weak, with scarce elements of visualization and transformative capacities. We want to show these problems by analyzing some elements of the normative pillars of institutionalization in Mexico and their weak embeddedness in social contexts. Anticipatory communities do not exist or do have low capacities of anticipation (González-Arellano 2025).

11.5 The Urban Plastic Economy in Mexico City as an Anticipatory System

The capacity to anticipate the consequences for the future and build more robust long-term strategies is crucial, as well as the main arena where plastic is used, as plastic pollution originates in the urban arena. We want to analyze the potential of these efforts, introducing the capacity to imagine and build less invasive futures in cities as anticipatory systems. In this section, we analyze the images of the future and the anticipatory assumptions related to the efforts to control and regulate the mitigation of plastic pollution in Mexico. To analyze the types of anticipatory approaches that have been coexisting, it is important to understand the tendencies in the national political context on these topics as reflected in federal and local regulations passed by Congress. Both arenas are important to understand the uses of future images in the buildup, application, and evaluation of these efforts.

During the last three decades, the interest of environmental groups and parliamentary groups in addressing the problem of plastic pollution brought 16 initiatives to the Chamber of Deputies and 22 to the Chamber of Senators, mainly focused on regulating bags and adding other single-use plastics in 2018 (WRI 2020). Since 2003, the main institutional channel for this regulation has been the General Law for the Prevention and Comprehensive Management of Waste (LGPGIR). The last modification of this law occurred on December 1, 2018, and gave rise to important guidelines related to plastic control. Article 7, for example, establishes that it is the Federation's responsibility to issue official **standards** for plastics. Article 28 states that both generators and producers, as well as importers, exporters, and distributors of plastic containers, including expanded polystyrene, as well as importers and distributors of used tires, are obliged to formulate and execute **plastic management plans** under the principles of valorization and **shared responsibility**. Our review of state legislation on environmental and waste management allowed us to identify that in the first three years of the new federal administration, between 2018 and 2020, 18 entities modified their environmental or waste management legislation to mention the management of plastics and bioplastic explicitly.

The current initiatives in Mexico are related to the political cycle that started with the victory of Andrés Manuel López Obrador and his MORENA movement in the presidential elections in 2018 and also in the state elections in México City.

The majority in the local parliament has allowed MORENA to push the approval of bills with the support of other political parties in their coalition (INE 2018, Oxford Analytica 2018). The announcement of Mexico City's new Solid Waste Law in 2019 represented a significant step in implementing a regulatory framework for reducing plastic pollution. It prohibited the selling and distributing of plastic bags since December 2020 and included other single-use plastics from January 2021 onward. The Mexican Ministry of Environment (SEMARNAT) produced in 2019 a policy document entitled "A National Vision towards Sustainable Management: Zero Waste" (SEMARNAT 2019) that embraced **zero waste** futures. It has been intended as a general guideline to transition toward a circular economy. The document establishes general **goals toward 2030**, related to specific images that the waste management policy has been aimed to reverse the traditional priorities that concentrated efforts in waste management, to transition to policies that prioritize prevention, minimization, and reuse, followed by alternatives such as secondary recycling, economic recovery, and final disposal.[8]

To put this vision into practice, actions related to monitoring, communication, education, co-responsibility of stakeholders, government coordination, and the promotion of citizen participation were proposed. These could, in principle, be operators of change in the anticipatory urban system, but for the moment, only their importance and statements are recognized in a very general way. However, incipient and subtle changes in plastic manufacturing and education campaigns have begun to be introduced due to national agreement among plastics producers and the Mexican Senate. They have placed their vision of the future exclusively on targeting the increase in PET collection and recycling content for 2025 and 2030 (Senado de la República 2019).

Two crucial regulatory documents were updated at the federal level as part of the strategy to implement this approach. On the one hand, the General Law for the Prevention and Comprehensive Management of Waste, and on the other, The National Program for the Prevention and Comprehensive Management of Special Management Waste. Both insist on the importance of generating **information systems** on waste's health and environmental effects, including plastics. Along these lines, one of the most relevant strategies and actions is formulating a **national inventory of sources of plastic pollution**. This inventory was presented in 2023 as part of the National Action Plan on Marine Waste and Plastic Pollution (SEMARNAT 2023). Additionally, the concept of shared responsibility, in which producers and users are involved, has been

[8] This policy document proposes expanding the coverage of collection services and infrastructure development, increasing the capacity for waste recovery, strengthening the separation of waste at source, promoting recycling, and establishing an effective inspection and surveillance body. The document proposes a general road map, which conceives six actions, among them: a comprehensive diagnosis (of waste, its management, infrastructure, and the regulatory framework); closure of final disposal destinations that do not comply with regulations; getting international expert advice to promote private investment for the development of technologies that allow the use and recovery of waste; local development of appropriate management models that benefit from best practices; transformation of open-air dumps into material banks; as well as avoiding food waste and taking advantage of its organic and energy potential (SEMARNAT 2019).

incorporated, intending to transcend a waste management approach and conceive it as a more systemic problem that involves producers.

Implementing the legal provisions for managing plastic and its waste was part of a broader program with a **circular approach**. The Government of Mexico City, through its Environment Secretariat (SEDEMA), presented the "Zero Waste Action Plan towards a Circular Economy" in May 2019. It is a set of actions aimed at reducing waste generation and no longer solely focused on waste management at the end of the product lifespan. The Plan has set goals for the short (2024) and medium term (2030). The Plan consists of actions aimed at separating and collecting waste, training and information campaigns, infrastructure development, and the dignity and recognition of the work of the cleaning system employees. In a transversal way, the issue of plastic crosses these lines of action in a very relevant way. Their promotional slogans mention: "A city with zero waste in 2030."

In response to the release of the governmental Zero Waste program, Greenpeace, along with three other relevant environmental organizations,[9] positioned itself by stating that "the 'National Vision of Sustainable Management: Zero Waste' does not solve the root problem," as it highlights waste management instead of waste reduction and overproduction. There is also concern that the official position does not ban the burning of waste, including plastics, for energy generation, enabling the release of toxic substances and greenhouse gases (Greenpeace 2019a, b). In this context, the burning of plastic and other waste is contemplated in the prospective scenarios of the cement industry to replace traditional petroleum coke. It is intended to increase these alternative fuels to 20% in 2025, 25% in 2030, and 35% in 2050 (Rappo 2020). A couple of years later, Marisa Jacott, representative of Common Frontiers, insisted during an interview that "There is no clear policy on the part of the government to implement a true zero waste policy" (Greenpeace 2021).

Mexico City has also been the center for environmental civil society organizations concerned about the excessive production of plastics and their poor management. In 2019, several of them integrated a broad network called "Mexico Zero Plastic Alliance." This network brings together recognized organizations such as Greenpeace but is open to other organizations and citizens. It currently includes 241 members and 125 organizations nationwide (https://alianzamexicosinplastico.org/). These organizations have been very skeptical regarding the scope of the legislation discussed above (Greenpeace 2024). They have criticized, for example, that the responsibility for the reduction of debris is still placed mainly on the side of the consumers and that it is very soft concerning limiting production or placing greater responsibility on the side of plastic producing and distributing companies (AMSP 2021).

Analyzing the structure of the urban anticipatory system, we can distinguish three anticipatory communities: plastics industrialists, a group of legislators and public officials, and environmental organizations. The established plastics industry, aware of public discontent and the incoming policy initiatives to regulate plastic production

[9] Center for Analysis and Action on Toxics and their Alternatives (CAATA), Common Borders, Community Development Research Laboratory (LIDC).

and consumption, has promoted campaigns to encourage recycling as a solution to the problem in the future. Likewise, it has sought to discredit the use of substitute materials as a feasible solution at the present. Policymakers and government officials have needed to legitimize their government programs in line with the current international vision of zero waste environmental policy and, at the same time, try to reconcile the competing interests of plastics industrialists and the pressure of local environmental organizations.

Concerning the regulatory scope, we identify that this is very varied and covers a broad spectrum of future transformations of the urban plastic economies ranging from (a) calls to encourage the use of alternative products, including biodegradable ones, (b) promote recycling, (c) sanction poor management of waste, and even (d) prohibit the marketing of bags and other single-use plastics. This change at the regulatory level did not immediately generate significant changes in production and consumption patterns due to the lack of complementary legal elements in lower levels (e.g., adequate norms and certifications) and to the slow changes in the availability of raw materials and alternative products, its technical standards, adequate surveillance and sanction mechanisms, among others.

We can observe that there has been a lack of visualization of alternative futures and also a low transformative capacity of guiding transition inside the communities of actors. It is also important to note that the health measures related to the COVID-19 pandemic relaxed the regulations. One of the central issues is the divergence in the future images between the pressures of business actors and environmental groups. Incumbent plastic producers have tried to place responsibility on public administration and consumers, suggesting the improvement of municipal collection systems and establishing recycling as the primary measure. At the same time, environmental groups and some academics insisted that responsibility must be shared with producers, promoting the substitution of materials, prohibiting unnecessary or single-use plastics, discouraging the burning of plastics to generate energy, and only opting for recycling when there is no other alternative (GAIA 2024).

11.6 Conclusions

Plastics have become an integral part of modern production and consumption systems with many important properties and applications that make replacing or reducing their usage difficult. This is another example of the rigidities in modern economies' technical, economic, cultural, institutional, and social configurations. Nevertheless, at the same time, the accelerated use of resources involved, particularly single-use plastics, has led to enormous resource waste and is causing significant risks, as plastic pollution is creating severe risks to human health and the environment. Moreover, these risks are only enhancing, as forecasts based on stable conditions show that the plastics economy will still grow substantially in the next decade.

The diverse representations of the plastics ecosystems and the diverse intervention proposals show a limited and strongly biased use of future images, narratives, and scenarios with frequently very narrow anticipatory assumptions. Individuals and society generally have a limited capacity to visualize futures and build solutions. However, the successful transition of systems of production and consumption of plastics toward less invasive or even sustainable models depends, among other things, on the capacity to create, compare, and apply long-term scenarios with diverse intervention options on different levels. Current futures studies offer more inclusive forms of revision, formulation, and application of alternative narratives. Anticipation theory and futures literacy promote individual, organizational, and social capacities of anticipation and pathways to multi-level governance.

Strong anticipation lies in the human conceptualization of uncertainty and complexity, the importance of an open and emergent construction of desirable futures, and the participatory and interactive governance of visualization and transformation processes. Anticipation theory revises mental models, imaginaries, and narratives of the future with novel analysis and practice tools. The first step in the participatory analysis of anticipatory systems is characterizing their images and the main processes of their creation, including mental models with all their biases and prejudices. Frequently, participating actors recognize the prevalence of low anticipatory capacities, linear anticipatory assumptions, and narrow pathways.

In our analysis of the urban plastic economy in Mexico City as an anticipatory system, we observed low capacities for visualization of futures for the plastic economy and very weak capacities for transformation. Many initiatives include images of short-term futures (related to the political cycle) that are fragmented and empty. Existing anticipatory communities cultivate futures narratives connected to basic concepts of plastic waste reduction but lack consistent anticipatory assumptions and governance mechanisms. For these reasons, the urban plastic economy in Mexico City has had a deficient transformative capacity.

We have also identified that the Mexican policy cycle allowed the reframing of customary waste programs narratives and that in such endeavor, urban anticipatory communities have played an essential role by introducing the international wording—even heterogeneous—about different expectations on a new plastics economy, but still struggle to translate alternative visions on new operative models to mature anticipatory urban capabilities.

References

AMSP (2021) Senadores simulan ley 'antiplásticos' para favorecer a la industria. https://alianzamexicosinplastico.org/wp-content/uploads/2021/10/Comunicado-de-prensa-AMSP-oct-2021.pdf

Andrady AL, Neal MA (2009) Applications and societal benefits of plastics. Philos Trans R Soc B Biol Sci 364:1977–1984

Barkela B, Werling K, Filler SE, Heidbreder LM (2021) Fasting plastic—the role of media reports in a 'window of opportunity' to reduce plastic consumption. Sustainability 13:11807. https://doi.org/10.3390/su132111807

Baskar et al (eds) (2022) Handbook of solid waste management. Sustainability through circular economy. Springer, Singapore

Borrelle SB, Ringma J, Law KL, Monnahan CC, Lebreton L, McGivern A, Murphy E, Jambeck J, Leonard GH, Hilleary MA, Eriksen M, Possingham HP, De Frond H, Gerber LR, Polidoro B, Tahir A, Bernard M, Mallos N, Barnes M, Rochman CM (2020) Predicted growth in plastic waste exceeds efforts to mitigate plastic pollution. Science 369(6510):1515–1518. https://doi.org/10.1126/science.aba3656

Brandao M, Lazarevic D, Göran F (eds) (2020) Handbook of the circular economy. Edward Elgar, Cheltenham

Dumbili E, Henderson L (2020) The challenge of plastic pollution in Nigeria. In: Plastic waste and recycling. Elsevier Academic Press, pp 569–583

Ellen McArthur Foundation (2021) Responsabilidad Extendida del Productor. Una parte necesaria de la solución para los residuos y la contaminación de los empaques y envases. https://plastics.ellenmacarthurfoundation.org/epr#Position-paper

EPA (United States Environmental Protection Agency) (2024) Zero-waste case study: San Francisco. https://www.epa.gov/transforming-waste-tool/zero-waste-case-study-san-francisco

European Commission: Directorate-General for Environment (2021) Turning the tide on single-use plastics. Publications Office of the European Union. https://data.europa.eu/doi/10.2779/800074

Fischer-Kowalski M, Haberl H (2007) Conceptualizing, observing and comparing socioecological transitions. In: Fischer-Kowalsk M, Haberl H (eds) Socioecological transitions and global change: trajectories of social metabolism and land use. Edward Elgar Publishing, pp 1–30. https://www.e-elgar.com/shop/usd/socioecological-transitions-and-global-change-9781847203403.html

GAIA (2022) La responsabilidad extendida del productor (REP). https://www.no-burn.org/wp-content/uploads/2022/06/UNEA-publication-packet_EPR_ES.pdf

GAIA (2024) Tratado Mundial sobre Plásticos: Menos Plásticos, Más Vida. https://www.no-burn.org/es/unea-plastics-treaty/. Access Feb 23

González-Arellano S (2025) Cities as anticipatory systems. In: González-Arellano S, Gandlgruber B (eds) Cities as anticipatory systems. Springer. https://link.springer.com/book/9783031781612

Greenpeace (2019a) Throwing away the future: how companies still have it wrong on plastic pollution 'solutions.' https://www.greenpeace.org/usa/research/how-companies-still-have-it-wrong-on-plastic-pollution-solutions/

Greenpeace (2019b) La visión nacional para gestionar los residuos no resuleve el problema de raíz. https://www.greenpeace.org/mexico/noticia/1726/la-vision-nacional-para-gestionar-los-residuos-no-resuelve-el-problema-de-raiz/. Access 5 Feb 2024

Greenpeace (January 16, 2021) Cuando la basura nos alcanzó y nos rebasó. https://www.youtube.com/watch?v=1aqC_xinDgo

Greenpeace (2024) ¡Libérate del Plástico! ¡Exijamos leyes que prohiban los plásticos de un solo uso!. https://www.greenpeace.org/mexico/participa/liberate-del-plastico/

Hannon J, Zaman AU (2018) Exploring the phenomenon of zero waste and future cities. Urban Sci 2(3):90

INE (2018) Cómputos de entidad federativa y circunscripción 2018. Elecciones Federales. https://computos2018-entidad.ine.mx/#/senadurias/nacional/1/2/1/2

Iroegbu AOC, Ray SS, Mbarane V, Bordado JC, Sardinha JP (2021) Plastic pollution: a perspective on matters arising: challenges and opportunities. ACS Omega 6(30):19343–19355. https://doi.org/10.1021/acsomega.1c02760

Jambeck JR, Geyer R, Wilcox C, Siegler TR, Perryman M, Andrady A, Narayan R, Law KL (2015) Marine pollution. Plastic waste inputs from land into the ocean. Science 347(6223):768–771. https://doi.org/10.1126/science.1260352

Kaza S, Yao L, Bhada-Tata P, Van Woerden F (2018) What a waste 2.0: a global snapshot of solid waste management to 2050. World Bank Publications. Urban Development Series. Washington, D.C. https://doi.org/10.1596/978-1-4648-1329-0

Kole PJ, Löhr AJ, Van Belleghem F, Ragas A (2017) Wear and tear of tyres: a stealthy source of microplastics in the environment. Int J Environ Res Public Health 14(10):1265. MDPI AG

Law KL, Starr N, Siegler TR, Jambeck JR, Mallos NJ, Leonard GH (2020) The United States' contribution of plastic waste to land and ocean. Sci Adv 6(44):eabd0288. https://doi.org/10.1126/sciadv.abd0288

Lebreton L, Andrady A (2019) Future scenarios of global plastic waste generation and disposal. Palgrave Commun 5(1):1–11

Lindhqvist T (2000) Extended producer responsibility in cleaner production: policy principle to promote environmental improvements of product systems. PhD Thesis, Lund University. https://portal.research.lu.se/en/publications/extended-producer-responsibility-in-cleaner-production-policy-pri

Mauch C (ed) (2016) A future without waste?: Zero-waste in theory and practice. Rachel Carson Center for Environment and Society

Mavropoulos A, Nilsen AW (2020) Industry 4.0 and circular economy: towards a wasteless future or a wasteful planet? Wiley

Miller R (2018) Transforming the future: anticipation in the 21st century. Taylor & Francis, p 300

OECD (2022a) Global plastics outlook: economic drivers, environmental impacts and policy options. https://www.oecd-ilibrary.org/environment/global-plastics-outlook_de747aef-en

OECD (2022b) Global plastics outlook: policy scenarios to 2060. OECD Publishing, Paris.https://doi.org/10.1787/aa1edf33-en

OECD (2023) Towards eliminating plastic pollution by 2040. A policy scenario analysis. Interim findings. https://www.oecd.org/environment/plastics/Interim-Findings-Towards-Eliminating-Plastic-Pollution-by-2040-Policy-Scenario-Analysis.pdf

Oxford Analytica (2018) Morena triumph changes the face of Mexican politics. Expert Briefings.https://doi.org/10.1108/OXAN-DB235869

Poli R (2019) Handbook of anticipation. Theoretical and applied aspects of the use in decision making. Springer

Pop V, Ozunu A, Petrescu DC, Stan AD, Petrescu-Mag RM (2023) The influence of media narratives on microplastics risk perception. PeerJ 11:e16338. https://doi.org/10.7717/peerj.16338

Prüss-Ustün A, Wolf J, Corvalán C, Bos R, Neira M (2016) Preventing disease through healthy environments: a global assessment of the burden of disease from environmental risks. World Health Organization, Geneva, Switzerland

Rappo S (February 28, 2020) Quemar residuos municipales e industrials en Cemex, parte de los mecanismos de mitigación al cambio climático. Jorn Oriente. https://www.lajornadadeoriente.com.mx/puebla/quemar-residuos-municipales-e-industriales-en-cemex-parte-de-los-mecanismos-de-mitigacion-al-cambio-climatico/

Schwartz JO (2023) Strategic foresight. An introductory guide to practice. Routledge. https://www.routledge.com/Strategic-Foresight-An-Introductory-Guide-to-Practice/Schwarz/p/book/9781032299235

SEMARNAT (2019) Visión nacional hacia una gestión sustentable: Cero residuos. https://www.gob.mx/cms/uploads/attachment/file/435917/Vision_Nacional_Cero_Residuos_6_FEB_2019.pdf

SEMARNAT (2023) Inventario Nacional de Fuentes de Contaminación Plástica

Senado de la República (2019) Acuerdo Nacional para la Nueva Economía del Plástico en México. https://cespedes.org.mx/wp-content/uploads/2021/02/Acuerdo-Nacional-de-la-Nueva-Economia-del-Plastico-en-Mexico-version-final-Senado.pdf

Teh PL, Piao S, Almansour M, Ong HF, Ahad A (2022) Analysis of popular social media topics regarding plastic pollution. Sustainability 14:1709. https://doi.org/10.3390/su14031709

Teles M, Balasch JC, Oliveira M, Sardans J, Peñuelas J (2020) Insights into nanoplastics effects on human health. Sci Bull 65(23):1966–1969. https://doi.org/10.1016/j.scib.2020.08.003

UN (2015) Transforming our world: the 2030 agenda for sustainable development. Resolution Adopted by the General Assembly on 25 Sep 2015, 42809, pp 1–13

UNEP (2018) Límites legales de los productos de plástico desechables y los microplásticos: examen a escala mundial de las leyes y los reglamentos nacionales. https://wedocs.unep.org/handle/20.500.1/1822/27113

UNEP (2020) Tackling plastic pollution: legislative guide for the regulation of single-use plastic products. https://wedocs.unep.org/handle/20.500.11822/34570?show=full
UNEP (2022) Draft resolution on an internationally legally binding instrument on plastic pollution. https://wedocs.unep.org/bitstream/handle/20.500.11822/37395/UNEA5.2%20Global_Agreement_Explanatory%20note%20and%20Resolution%2027%20October.pdf?sequence=1&isAllowed=y
UrbanShift (2023a) Foro Internacional de Paris para Acabar con la Contaminación por Plásticos en las Ciudades. https://es.shiftcities.org/event/paris-international-forum-end-plastic-pollution-cities
UrbanShift (2023b) Cómo las Ciudades Lideran la Lucha contra la Contaminación por Plásticos. https://es.shiftcities.org/post/how-cities-are-leading-fight-against-plastic-pollution
WRI (2020) Mecanismos de política para disminuir los residuos plásticos de un solo uso: Revisión de las opciones disponibles y su aplicabilidad en México. https://wrimexico.org/publication/mecanismos-de-pol%C3%ADtica-para-disminuir-los-residuos-pl%C3%A1sticos-de-un-solo-uso#:~:text=%C3%89stas%20describen%20el%20estado%20actual,4)%20responsabilidad%20extendida%20del%20productor
Zero Waste International Alliance (ZWIA) (February 19, 2019) Working towards a world without waste. https://zwia.org/
Zero Waste International Alliance (February 23, 2024) Zero waste definition. https://zwia.org/zero-waste-definition/
Zero Waste Revolution (February 23, 2024) Our mission: zero waste. https://zerowasterevolution.net/

Part V
Anticipatory Elements for Urban Climate Policies

Chapter 12
Recife and Anticipation of Extreme Weather Events: Paths to Advance in Climate Justice

Pedro Roberto Jacobi, Marcos Tavares de Arruda Filho, and Rafael Calderón-Contreras

Abstract Anticipatory governance represents the need for changes in a long-term vision based on multiple strategies that prevent future crises, especially those arising from the global environmental emergency. In Brazil, around 80% of people already live in urban centers with multiple dimensions of vulnerability, where the issue of anticipation and inclusion in local government agendas is fundamental for identifying and implementing initiatives for reducing future impacts. Recife, located in the state of Pernambuco, is the 16th city most threatened by climate change in the world due to its low altitude and the large number of watercourses across its territory. This work aims at discussing the anticipatory nature of Recife's urban policies to prevent further extreme climate disasters that already affect the city. The presence of a framework of public policies that form an anticipatory system for extreme weather events means that coastal cities like Recife may open a wide array of possibilities for reducing the risks and impacts associated with climate change-related crisis. Anticipatory systems rely on the partnership between municipal institutions and other cooperation networks for inserting debates such as climate justice, which are sometimes limited to just the academic sphere within the decision-making arena.

Keywords Anticipatory systems · Socio-ecological transition · Cities · Recife · Climate justice

---

P. R. Jacobi · M. T. de Arruda Filho
Instituto de Energia E Ambiente—Universidade de São Paulo, Recife, Brasil
e-mail: marcostavares@usp.br

R. Calderón-Contreras (✉)
Department of Social Sciences, Universidad Autónoma Metropolitana-Cuajimalpa, Mexico City, Mexico
e-mail: rcalderon@cua.uam.mx

S. Gonzalez-Arellano and B. Gandlgruber (eds.), *Cities as Anticipatory Systems*, The Urban Book Series, https://doi.org/10.1007/978-3-031-78162-9_12

12.1 Introduction

Debates on local governments and climate change, especially in the urban context, have become fundamental in several areas, such as public management. Anticipatory governance represents the need for changes in a long-term vision based on multiple strategies that prevent future crises, such as the one arising from the global climate issue. In Brazil, around 80% of people already live in urban centers. Cities increasingly demand more attention to climate change due to their high vulnerability to the effects of climate change worldwide.

In the context of Northeast Brazil, where populations and territories are vulnerable in multiple dimensions, including climate change, the issue of anticipation and inclusion in local government agendas is fundamental to identifying and implementing initiatives to reduce their impacts. Most state capitals in the region have climate adaptation plans, mainly due to the exposure of these cities to rising sea levels and their location close to the coast.

Recife, located in the state of Pernambuco, is the 16th most threatened city by climate change in the world due to its low altitude and the large number of watercourses that cross its territory. This article aims to show how, in the case of the city of Recife, there is some level of internalization of the theme of climate change in municipal urban policy and the challenges that exist in deepening actions that present the scope of public policies that in collaboration with cooperation networks such as ICLEI, are configured as a climate governance arrangement, whose primary role is to contribute to promoting local actions to combat the climate crisis.

This work aims to discuss the anticipatory system within urban policies focused on climate issues in Recife to prevent extreme climate disasters that already affect the city. The text is divided into five sections: this introduction addresses the theme analyzed and raises the main topics to be discussed. The second section aims to frame the importance of anticipatory approaches to dealing with the climatic transition in cities. The Socio-ecological Transition in Cities discusses how cities prepare for a new ecological paradigm. The following section addresses the Anticipatory Climate Governance in Brazil, which provides a history and a discussion on how multilateral governance has been working on the climate issue in the country. The fourth section deals with the case of the Brazilian Northeast: Signs of Change in the Climate Transition, discussing how the states in this region address climate change within their public policies. The chapter then discusses the issue of Climate Justice in Recife: Transition to Anticipation, which brings the evolution and construction of a history of public policies focused on the climate and the final considerations in which there is closure and direction so that more cities can work on anticipatory systems for global climate change.

12.2 The Social-ecological Transition in Cities

As mentioned elsewhere in this book, cities are social-ecological systems entitled to permanent shifts and complex dynamics of change. Some of those changes are based on a disruption of some of the components of these systems. Global environmental change has become one of such transgressive factors that originate profound transitions in the structure of cities as systems. Global environmental change has associated risks that pose severe vulnerabilities to city populations and their biophysical grounds. Furthermore, those risks also negatively impact the social, political, and economic traits that constitute cities' social organization and governance of cities worldwide. Vulnerability has become one of the normalized states in which cities develop and grow. However, a growth process that often does not include a formalized process of anticipation translated into a general reduction of risks associated with global environmental change.

Anticipation in the context of cities' social-ecological transition is a necessary goal for governments and social institutions alike. As this chapter illustrates, the city of Recife has implemented a series of mechanisms to deal with scenarios related to the future conditions expected due to climate change. As analyzed by González, in this volume, imagining the future is one of two dimensions of anticipation that cities as anticipatory systems must implement to deal with the future consequences of external risks. Given the vast array of threats that climate change implies for Recife, considering the future is one of the critical aspects that public policies and plans for facing climate change must consider.

The second dimension of anticipation in cities as social-ecological systems is the possibility of incorporating the future into the system's transformation. In light of the ever-increasing challenges associated with climate change, cities, like any social-ecological system, react to different stimuli to change specific traits at determined scales. These changes are associated with transitions, which are chained to other transitions in different system qualities and may lead to social-ecological transformations that may lead to a better state of resilience in the system (Sievers-Glotzbach and Tschersich 2019). In the case of cities, this chain of transitions may imply the construction of social-ecological resilience to climate change and its impacts.

Building social-ecological resilience to withstand the effects of climate change in cities requires a process of institutionalization of anticipation (González-Arellano 2022). This process of institutionalization may often depart from the urban governance derived from a wide array of policies designed. The case of Brazil is illustrative when it comes to this effort. From national policies to adopting and adapting such policies in urban contexts, the potential for anticipating the conditions that cities will face due to climate change is paramount for creating resilience. As this book specifies, the regulatory pillar for which it is possible to enforce future scenarios in cities must be derived from multi-scalar governance, but also on the local actor's ability to foresee the impacts of climate change. This chapter uses the case of Recife to illustrate how the multi-scale governance in Brazil derives into an anticipatory system with the potential for building resilience in the face of climate change.

The transition of Recife, as a social-ecological system and a more resilient city, implies a wide array of anticipatory processes. Such processes characterize the social-ecological transition that cities must endure to promote more resilient futures. The case of Recife is illustrative of a factor often denied in the analysis of social-ecological transitions in cities, and it is the need for justice as the core for the construction of imaginaries of the city's future. Justice becomes not only the final goal of a process of anticipatory governance but the process by which a city, understood as a social-ecological system, may rely on its regulatory and normative governance structure and biophysical structure. Combining these traits (the ecological and the social) provides the means for creating anticipatory systems that may derive a better future for cities: one that appeals to a safe and just social-ecological system.

12.3 Anticipatory Climate Governance in Brazil

Actions aimed at environmental issues, according to the Federal Constitution, the National Environmental Policy—PNMA (1981), and the National Adaptation Plan—PNA (2016), require integration of policies considering the different levels of government, sectors and a diverse set of territorial characteristics that go beyond the borders of states and municipalities. In large urban agglomerations, the environmental governance of urban space must be innovative to face social pressures in the context of urbanization and the vulnerabilities of populations associated with occupation in areas at risk of landslides or subject to flooding. These are risks associated with poverty and inequality in urban logic, exposing the lack of urban planning and, notably, the use and occupation of land. However, when addressing the issue of climate change, it becomes necessary to address the risks; as Beck (2010) states, disasters are announced and cannot be seen as fatalities and can, in most cases, be predicted and avoided.

The main aspects (variables) that directly or indirectly influence the climate adaptation capacity of cities are systematized by studies such as that by Teixeira et al. (2020). The authors emphasize access and use of information; capacity to allocate economic, institutional, and technological resources; the intersection between adaptation, mitigation, and sustainable development; synergies between generic and specific capabilities; share capital; trust relationships; presence of public policy entrepreneurs; and participatory governance processes. Risk reduction and adaptation management is a transversal theme, and this demands that city formulators and planners understand how climate adaptation needs to be incorporated or integrated into urban planning practices (ICLEI, CEPS, 2013).

At the federal level, the central normative frameworks are the National Climate Change Policy of 2009, the National Civil Protection and Defense Policy in 2012, and the National Adaptation Plan in 2026. The PMMC defines in its Article 5 the objective of contributing to stabilizing greenhouse gases in the atmosphere "within a sufficient time to allow ecosystems to adapt naturally to climate change." It determines its implementation in sectoral axes, and adaptation focuses on health and land use axes, emphasizing the elimination or minimization of risks and adaptation measures related to climate change. Its guidelines include coordinated action between the Union, States, and Municipalities, with the Union coordinating and structuring policy instruments and supporting other entities in mapping risk areas and prevention and response actions, as well as the obligation to transfer resources to them in case of disaster.

The National Plan for Adaptation to Climate Change (PNA), established in 2016, presents the guideline that adaptation to climate change includes the improvement of urban planning models, integrating the planning of land use and occupation with environmental preservation to prevent the risks of natural disasters (CAF/WAYCARBON/ICLEI 2019). It is observed that the occurrence of extreme events, such as water scarcity, exposes the authorities' lack of preparation to deal with these issues and the lack of a culture of prevention in society. Preventive management of climate risks is still more in theory than practical actions (Jacobi and Sulaiman 2016). It is worth noting that many municipalities still do not have an adequate civil defense structure for preventive risk management. In the face of uncertainty and multiple risks, this requires that the preventive management of climate risks and the promotion of actions guided by a sustainability agenda are highlighted, thus reducing vulnerability with innovative management and urban planning. In this sense, adaptation measures can be adopted to reduce socio-environmental costs and losses in vulnerable social groups.

In 2016, in an effort led by the federal government that included the collaboration of civil society, the private sector, and state governments, the National Plan for Adaptation to Climate Change (PNA) was established. Its main objective is to promote the management and reduction of climate risk in the country in the face of the adverse effects of climate change, to take advantage of emerging opportunities, avoid losses and damages, and build instruments that allow the adaptation of natural, human, production, and infrastructure.

In national communications (2016), an index of vulnerability to landslide and flood disasters was created. Subsequently, a second indicator was created on vulnerability to droughts, information that supported the PNA (Camarinha 2019). These indicators make the relationship between exposure and sensitivity (or susceptibility), resulting in potential impacts. To this end, a map of potential impacts was prepared in the current situation, and estimates for 2040 to estimate future impacts.

Brazil's Fourth National Communication to the United Nations Framework Convention on Climate Change (UNFCCC), sent to the UNFCC in December 2020, assesses vulnerability, impacts, and adaptation measures in an integrated manner (Nexus + approach), considering water security, energy, food and socio-environmental (Coutinho et al. 2019). In the Third Communication in 2016, an index of vulnerability to landslide and flood disasters was created, and subsequently, a second indicator was created on vulnerability to droughts, information that subsidized the PNA (Camarinha 2019).

The IPCC proposes new emission scenarios (IPCC 2022) that allow new projections of global climate change to be defined, with global Earth system models becoming the main tools for studies. Based on this innovation, several Brazilian government institutions that develop research in science and technology in convergence with global models from England and Japan have developed regionalized models for the Brazilian territory with a national model developed by the National Institute for Space Research-INPE (Chou et al. 2014). The indicators mentioned above make it possible to verify potential impacts based on the mapping of possible actions that municipal governments could implement to mitigate future impacts.

Among the problems observed, some of the most pressing ones are faced by municipal and regional governments with measures based on how infrastructure is being used for reducing risks. Urban infrastructure is often directed to reducing risks caused by heat islands reduced air quality and accelerated stormwater runoff. Therefore, municipal administrations, being responsible for implementing adaptation to the effects of climate change, must promote changes in the governance logic, stimulating intersectoral actions; promoting housing policies for the most vulnerable sectors and strengthening preventive and transversal actions incorporating adaptive responses to climate change in master plans and land use and occupation legislation.

Romero-Lankao and Gnatz (2019) highlight the importance of incorporating equality in the definition of urban adaptation actions, as inequality in exposure to risks shapes populations' mitigation and adaptation capacity. In this sense, there are no possible goals for plans or policies related to climate justice without local co-creation work. On the contrary, what is suggested is an inversion in how plans and policies are constructed with the inclusion of local actors, their training to influence these processes autonomously, and the structural themes described above, such as structural racism, among others.

Research carried out in the context of the *Latino Adapta project* with decision-makers related to adaptation to climate change systematized a set of aspects that require concrete actions and advances regarding (i) investments directed to research, maintenance of instruments for monitoring climate variables, technological advances, facilitation of the production process and dissemination of knowledge, among others; (ii) efforts to stimulate dialogue, raise awareness of the topic and translate data into applicable information; and (iii) promoting partnerships and synergies between actors and activity coordinators, to better use of information, as well as other resources, at all stages of the process of developing and implementing public policies. Finally, it is concluded that adaptation to climate change needs to be closely linked to local development. Identifying impacts and vulnerabilities and evaluating the various adaptation options must be part of the public manager's routine. Thus, the knowledge produced can clarify any measure selected based on the scenarios presented to generate benefits (with or without climate change) and avoid poor investments and actions that cause regret (Jacobi et al. 2019).

Osório (2016) points out that concerning the PNA, numerous challenges are to be faced and reinforces that the adaptation process demands enormous commitment from national leaders. Concerning municipal plans and strategies, Margulis (2017) states that the main challenges are in the planning and executing adaptation and mitigation strategies and that they are not yet prepared for the occurrence of disasters resulting from events related to climate change.

Multiple forms of environmental degradation occur, predominantly where lower-income populations, black communities, and indigenous groups live. There is a need to address issues of gender, social inequalities, and human rights together with social inclusion strategies and integrative responses that promote dialogue and reciprocity. This capillarity is also essential when implementing adaptation and loss and damage policies, as NGOs often already operate at the local level and in communities where public authorities are often not present. This creates a rapprochement with the population (Lopez and Abreu 2014). However, it is also important to highlight that an unequal distribution of resources makes it challenging to promote easy adaptation (Travassos et al. 2021).

To this end, the development of vulnerability indices must include the engagement of directly affected populations, facilitating access to information that enables their more effective participation; as stated by Teixeira et al. (2020), "municipal adaptation plans and strategies need to define parameters for decision-making at different adaptive scales: social vulnerability, urban development standards, and infrastructure, urban planning, and delimit which regions, sectors, areas and populations of the municipality should have their adaptive capacity increased to reduce future impacts related to projected climatic and meteorological events, and indicate which sectoral public policies should increase this adaptive capacity."

12.4 Brazilian Northeast: Signs of Change in the Climate Transition

Brazil presents an exclusion of the theme focused on the climate emergency from the decision-making process, making it difficult to debate the inclusion of the human element within the crisis. However, social concern about the issue is latent and urgent. This political lack of interest in the agenda reflects the scarcity of specific plans for adapting to climate change in cities and the growth of socio-climatically vulnerable communities. Local adaptation policies are fundamental in reducing socio-environmental vulnerability in each part of the country since the impacts of climate change are different for each location (Barbi and Rei 2021). However, as Coutinho et al. (2021) analyzed, Brazilian municipal adaptation plans and strategies need to define parameters for decision-making at different adaptive scales.

In the country's northeast region, Teixeira et al. (2020), when analyzing policies aimed at climate issues in the region's capitals, detected that only the cities of Salvador and Teresina incorporated concerns about climate issues into their municipal master plans. Fortaleza and Recife, capitals of the states of Ceará and Pernambuco, respectively, have strong integration of climate change into their government agendas, with the capital of Pernambuco playing a leading role in the climate emergency in the northeast and having an action policy considered successful for climate prevention and adaptation. (Oliveira et al. 2014; Debortoli et al. 2017). In this deficit of an anticipatory system of public policies, the greatest weakness of the climate transition in the northeast region of Brazil lies. The lack of policies focused on the social agenda within the climate issue exposes all previous social vulnerability aggravated by climate change when faced with an extreme climate event.

In another direction, cities on the coast of Brazil, mainly due to their exposure to rising sea levels and the fact that they are already feeling this effect along the entire coast, are more robust in the construction of public policies focused on the climate. Moraes (2020) considers Recife, one of the cities in this situation, as the city most sensitive to climate events and the triggering of disasters, which, consequently, presents this character of anticipation of upcoming events.

12.5 Climate Justice in Recife: Transition to Anticipation

Although extreme weather events affect the entire population, they do not affect everyone similarly. Different social groups are more exposed to its effects and consequences and have a lower capacity for resilience and restructuring immediately after the catastrophe. The inclusion of social justice in the climate change agenda was called climate justice. According to Sultana (2021), climate justice is fundamentally about paying attention to how climate change affects people differently, unequally, and disproportionately, as well as redressing the resulting injustices in fair and equitable ways. Climate justice means different things to different groups and individuals, and the potential implications of climate change for justice are varied and complex. It is not simply a quick reference for acting ethically but rather a source of reasoning for what is considered legitimate, which relates to more normative issues of equity and justice (Huntjens and Zhang 2016).

Recife is currently the Brazilian capital most exposed to extreme weather events; and due to its low altitude and also the large number of rivers and waterways, the 16th most threatened city by rising sea levels in the world (IPCC 2022) (See Fig. 12.1). The city comprises more than 100 rivers and streams that make up the hydrographic basins of the Capibaribe, Beberibe, and Tejipió rivers. This number of bodies of water and the composition of islands and bridges interconnecting them make the city known as "Brazilian Venice."

Recife's geography is compounded by poor urban planning, directly affecting low-income populations. The lack of adequate public policies that prevent residents from the consequences of rain means that the city is exposed to flooding when rainfall increases. Hill regions and flooded areas that comprise the city's relief suffer most from extreme weather events. The biggest climate catastrophe recorded in recent years occurred in May 2022, when it rained 679.1 mm, double the amount predicted for the same period (APAC 2022), and led to the death of 120 people from poor neighborhoods and hilly areas of the city.

Another climate event that worries the city is the advance of the sea over the coast. According to the Brazilian Network for Research on Global Climate Change (Rede Clima 2022), the sea has advanced an average of 0.54 cm from the coast

Fig. 12.1 Hydrography of Recife, Pernambuco, Brazil. *Scale* 1000 m *Source* Recife Data, *Google Maps, 2022*

of Recife per year since 1940. The city's coastline is already 45.7% of its length as having high vulnerability to advance and, according to the Climate Risk and Vulnerability Analysis, 81.8% of urban constructions that are less than 30 m from the coastline and on land below 5 m will be directly affected by changes in the water level (Recife 2023). PLAC-Recife (2020) states that places floods, landslides, heat waves, communicable diseases, meteorological drought, and rising sea levels are the leading climate threats to the city.

This problem is aggravated by the urban occupation of the city's coastline, which means that the sand strip is reduced yearly (Figs. 12.2 and 12.3). At the end of the 70s and beginning of the 80s, major anthropogenic changes occurred on Boa Viagem beach, the city's most famous tourist spot and one of the main urban beaches in Brazil. The construction of a seaside avenue (lanes, sidewalks, walls, kiosks, bathrooms, and leisure infrastructure) had a strong influence on the waterproofing of the land and immobilization of the dunes, which could have led to a severe displacement of the fragile sedimentary balance towards the erosion side of the beach. Oliveira et al. (2014) warn against limiting the future use of areas on Boa Viagem beach, as, due to the socioeconomic pressure occurring in the state of Pernambuco, the construction of large-scale works increases erosion at the site and produces a deficit in sand sedimentation, with a risk of reducing the coastline.

The fact that the region is in a privileged location in Recife means that the communities that inhabit Boa Viagem beach, together with their Special Zones of Social Interest (ZEIS), are targets of intense expectations of changes brought about by real estate and commercial appropriation (De Souza et al. 2020).

Historically, the municipal management of the city of Recife has been concerned about climate change and extreme events in its public environmental policies in

Fig. 12.2 Aerial photograph of the Recife waterfront in 1930. *Source* Joaquim Nabuco Foundation-FUNDAJ, 2022

Fig. 12.3 Aerial photograph of the Boa Viagem waterfront in 2021. *Source* G1 Pernambuco, 2022

recent years. In 2009, the Municipal Environment Fund-FMMA (Law No. 16,047) was created, which guarantees resources for the operation of the Secretariat for the Environment and Sustainability (Smas) and projects in the environmental and climate area. Then, in 2013, the Recife Sustainability and Climate Change

Committee—CONCLIMA and the Sustainability and Climate Change Group—GECLIMA (Decree No. 27,343/2013) were created, whose mission is to establish policies and actions to combat the climate emergency locally. CONCLIMA comprises six municipal departments and is coordinated by the Secretariat for Sustainable Development and Environment. In 2014, the Recife Sustainability and Climate Change Policy was created (Law No. 18,011 of 2014). The following year, the first inventory of greenhouse gas emissions in Recife was carried out.

In 2019, the Municipality of Recife's Analysis of Climate Risks and Vulnerabilities and Adaptation Strategy was launched in partnership with the City of Recife and ICLEI—Local Governments for Sustainability. This document is considered a critical turning point in the anticipatory management of extreme weather events in the city, as the risk is internalized for the first time, and the human and social component begins to be addressed within the climate issue.

The most recent of the public policies is the Local Climate Action Plan of the city of Recife (PLAC-Recife), which was created again by the partnership between Recife City Hall and ICLEI and internalizes and prepares the city for future extreme climate-related events; especially extreme rain and the consequences of rising sea levels. Being the result of a participatory process and bringing actions aimed at marginalized and vulnerable population groups. This document has climate justice as one of its basic principles, being one of the first Brazilian public policies to include this concept within its text. One of its characteristics is to address an anticipatory perspective on the current problems arising from climate change in the city through the axes of Energy, Sanitation, Mobility, and Resilience.

The presence of a framework of public policies that form an anticipatory system for extreme weather events means that coastal cities like Recife can assess vulnerable regions and parts of the population that may suffer the consequences in advance. That is important in decision-making and constructing urban infrastructure works for climate adaptation.

A specific anticipatory system for the city of Recife must permeate a predictive model about the main problems related to precipitation, especially during its period of greatest concentration, which has the greatest impact on the city. The social element must be at the heart of this debate since, with the inequality of resources that is present in this territory, the promotion of climate justice within this sphere will only happen by looking at the poorer sections of the population and their losses and damages through the growing extreme weather events in the city.

The existing plans within Recife's current public climate policy still have gaps regarding the social issue of climate change. The construction of anticipatory climate governance is necessary to strengthen climate justice in society and the state's practical actions towards those most affected. A strengthened anticipatory system for the city could take the form, for example, of a specific plan for hill areas and communities living close to the coast and riverbanks.

12.6 Final Considerations

Recife, like other cities with high climate vulnerability in the world, ended up emerging as an example in the formulation of an anticipatory system for extreme climatic events through the formulation of an extended framework of public policies focused on the issue and the promotion of an intense debate around them, with a strong insertion of the agenda within society.

The partnership between municipal management and cooperation networks such as ICLEI is essential in inserting debates such as climate justice, which are sometimes limited to just the academic sphere within the decision-making arena.

The challenges in metropolitan regions require public managers to implement policies and planning that consider social inequalities and climate change to ensure the quality of life for their socially vulnerable residents. Highlights include the expansion of urban resilience through the implementation of programs to recover environmental functions based on ecosystems, changing patterns of land use and occupation, reviewing construction standards and construction codes, implementing and expanding green infrastructure, improving air quality and urban mobility, strengthening the resilience of infrastructure and the universalization and qualification of sanitation services. However, local governments are also required to reinforce preventive management to deal with climate risks and promote adaptation measures considering social and environmental vulnerabilities, with transparency and social participation, emphasizing actions guided by climate justice. This demands management practices that consider strengthening dialogues essential for implementing adaptation policies. The role of NGOs that already operate in the local sphere and in communities where public authorities are often not present strengthens capillarity and rapprochement with the population for the elaboration of participatory diagnoses and for the execution of projects in a more horizontal and locally legitimized way.

This requires thinking about the climate crisis and adopting new practices in a context of networked, transversal, and democratic governance with interdisciplinary and aggregating approaches that incorporate into climate governance and intersectional and transformational policies that should be included in plans, policies, and programs for climate justice.

References

APAC. Agência Pernambucana de Águas e Clima (2022) https://www.apac.pe.gov.br/. Accessed on 2 Feb 2023

Barbi F, Rei FCF (2021) Mudanças climáticas e agenda de adaptação nas cidades brasileiras. Rev Catalana Dret Ambient 12(1)

Beck U (2010) Sociedade de risco. São Paulo: Ed 34:49–53

CAF/WAYCARBON/ICLEI (2019) Análise de riscos e vulnerabilidades climáticas e estratégia de adaptação do município do Recife—PE—Resumo para tomadores de decisão. http://www2.recife.pe.gov.br/sites/default/files/sumario_clima_recife_portugues.pdf

Camarinha P (2019) Status e Limites do Conhecimento em Mudança do Clima no Nível Municipal. Índice de Vulnerabilidade dos Municípios Brasileiros. CEMADEN (PPT)

Chou SC et al (2014) Evaluation of the eta simulations nested in three global climate models. Am J Clim Chang 3:438–454

Coutinho SM, Santos DV, Bursztyn M, Marengo JA, Lucena AF, Andres D, Maia SM (2019) A adaptação sob uma perspectiva integrada no Brasil: abordagem Nexus+. Rev Sustentabilidade em Debate, Brasília. https://periodicos.unb.br/index.php/sust/issue/view/2174/520

Coutinho SM, Maglio IC, carbone AS, Neder EA (2021) Adaptação às mudanças climáticas no Brasil: complexidade, incertezas e estratégias existentes. ClimaCom—Coexistências e cocriações [Online], Campinas, Year 8, Num. 20, April, 2021. https://climacom.mudancasclimaticas.net.br/adaptacao-as-mudancas-climaticas/

Debortoli NS, Camarinha PIM, Marengo J, Rodrigues RR (2017) An index of Brazil's vulnerability to expected increases in natural flash flooding and landslide disasters in the context of climate change. Nat Hazards

De Souza TO et al (2020) Bibliometria das publicações da revista 'Saúde em Debate': uma fotografia da saúde mental. Saúde Debate [Online] 44:305–319. spe 3, Accessed on 1 June 2022. https://doi.org/10.1590/0103-11042020E325. Epub: 12 Aug 2021. ISSN 2358-2898

González Arellano S (2022) The future scenarios of cities: an analysis of their institutional construction. Int J Urban Reg Res 46(5):865–873. https://doi.org/10.1111/1468-2427.13103

Google (s.f) Hydrography of Recife, Pernambuco, Brazil. Accessed on 15 Mar 2022

Huntjens P, Zhang T (2016) Climate justice: equitable and inclusive governance of climate action. Hague Inst Global Justice 1(1):1–29

ICLEI—Local Governments for Sustainability; CEPS—Centre for European Policy Studies (2013) Climate change adaptation: empowerment of local and regional authorities, with a focus on their involvement in monitoring and policy design. https://cor.europa.eu/en/documentation/studies/Documents/climate-change-adaptation.pdf. Accessed on 18 Oct 2018

IPCC (2022) AR6 climate change 2022: impacts, adaptation and vulnerability. impacts, adaptation and vulnerability. https://www.ipcc.ch/report/sixth-assessment-report-working-group-3/. Accessed on 11 Apr 2022

Jacobi PR, Sulaiman SN (2016) Governança ambiental urbana em face das mudanças climáticas. Rev USP 109:133–142

Jacobi et al (2019) Lacunas de conhecimento em adaptação às mudanças climáticas. Relatório Diagnóstico Brasil. Red Regional de Cambio Climático y Toma de Decisiones. Programa UNITWIN de UNESCO, Proyecto LatinoAdapta. https://saras-institute.org/wp-content/uploads/2020/05/Informe-Brasil_2019.pdf

Lopez FG, Abreu RA (2014) Participação das ongs nas políticas públicas: o ponto de vista de gestores federais. IPEA—Instituto de Pesquisa Econômica Aplicada—1949. Text for discussion, Brasília, April 2014. https://sbsa.com.br/wp-content/uploads/2020/08/Estudo-a-participa%C3%A7%C3%A3o-das-ONGS.pdf. Accessed on 20 Sep 2022

Margulis S (2017) Guia de Adaptação às Mudanças do Clima para Entes Federativos. Relatório completo WWF, Brasília-DF

de Moraes DA (2020) REVISITANDO AS ZONAS ESPECIAIS DE INTERESSE SOCIAL NO RECIFE: LIMITES E CONTRADIÇÕES PARA A GESTÃOE REGULAÇÃO DE FAVELAS. Oculum Ensaios 17

Oliveira DD, Souza-Santos LP, Silva HKP, Macedo SJ (2014) Toxicity of sediments from a mangrove forest patch in an urban area in Pernambuco (Brazil). Ecotoxicol Environ Saf 104:373–378. https://doi.org/10.1016/j.ecoenv.2014.02.004

Osorio G (2016) Planejamento Público: Desafios para a integração do tema adaptação. Rev Adaptação Mudança Clima/Ministério Meio Ambiente 1(1), June, 2016, Brasília: MMA

Recife, Prefeitura do (2023) Prefeitura do Recife. https://www2.recife.pe.gov.br/. Accessed on 16 Feb 2023

Rede Clima (2022) Mudanças Climáticas Globais no Brasil. http://redeclima.ccst.inpe.br/. Accessed on 2 Feb 2023

Romero-Lankao P, Gnatz D (2019) Risk inequality and the food-energy-water (FEW) nexus: a study of 43 city adaptation plans. Front Sociol 4
Sievers-Glotzbach S, Tschersich J (2019) Overcoming the process-structure divide in conceptions of social-ecological transformation: assessing the transformative character and impact of change processes. Ecol Econ 164(106361). https://doi.org/10.1016/j.ecolecon.2019.106361
Sultana F (2021) Critical climate justice. Geogr J [S.L.] 188(1):118–124. https://doi.org/10.1111/geoj.12417
Teixeira RLP, Pessoa ZS, Di Giulio GM (2020) Cidades, mudanças climáticas e adaptação: um estudo de caso de Natal/RN, Brasil. Desenvolvimento Meio Ambiente 54:468–483
Travassos L et al (2021) Why do extreme events still kill in the São Paulo Macro Metropolis Region? Chronicle of a death foretold in the global south. Int J Urban Sustain Dev 13(1):1–16

Chapter 13
Anticipatory Action and Future Urban Living in a Context of Increasing Temperatures: An Analysis from the Barcelona Climate Shelter Network

Brais Estévez, Julia Hidalgo, and Marion Bonhomme

Abstract In recent years, the world has undergone a significant transformation from a realm of certainties to one filled with uncertainties and looming disasters. These catastrophes, which are both ecological and social, are primarily attributed to the climate crisis caused by human activities. As a result, the future of cities and the planet is at stake, with the prospect of becoming uninhabitable. To address this urgent issue, local governments worldwide have taken action to mitigate and adapt to climate change. One exemplary initiative is the Barcelona City Council's Network of Climate Shelters, which was established in 2019. This network aims to prepare the city for the anticipated rise in temperatures by transforming existing public spaces, such as libraries and schools, into climate shelters. What sets the Barcelona Network of Climate Shelters apart, at least in one of its modes of assemblage, is its collaborative approach. The citizens actively participate in designing and implementing some of these infrastructures. That ensures the shelters are efficient and effective and fosters a sense of collective responsibility toward the climate crisis. By involving the public, the initiative prepares the city for climate change and builds a more resilient citizenry capable of facing an uncertain future together.

B. Estévez (✉)
Departamento de Xeografía, Universidade de Santiago de Compostela, Praza da Universidade, 1, 15703 Santiago de Compostela, Spain
e-mail: brais.estevez.villarino@usc.es

J. Hidalgo
National Centre for Scientific Research (CNRS), Laboratoire Interdisciplinaire Solidarités, Societés, Territoires (LISST), Toulouse II University, Toulouse, France
e-mail: julia.hidalgo@univ-tlse2.fr

M. Bonhomme
National Institute on Applied Sciences (INSA), Laboratoire Matériaux et Durabilité des Constructions (LMDC), Toulouse III University Toulouse, Toulouse, France
e-mail: marion.bonhomme@insa-toulouse.fr

S. Gonzalez-Arellano and B. Gandlgruber (eds.), *Cities as Anticipatory Systems*, The Urban Book Series, https://doi.org/10.1007/978-3-031-78162-9_13

Keywords Anthropocene · Cities · Climate shelters · Climate change adaptation · Heatwaves

13.1 Introduction: Why Do We Need to Talk About Climate Shelters?

Global warming is an unequivocally anthropogenic phenomenon that has triggered widespread and irremediable impacts that threaten life-support ecological systems and human and nonhuman well-being (Arias et al. 2021). In this sense, the IPCC reports are increasingly transparent about the direct link between global warming and the increase in extreme weather events such as heat waves, heavy precipitation, floods, droughts, wildfires, and others. The accumulation of scientific evidence on climate change and the awareness of its damaging effects has led governments and policymakers worldwide to include climate change mitigation and adaptation projects in their agendas—especially since the Paris Agreement at COP21 in 2015.

However, despite scientific evidence alerting to the risks for the viability of contemporary human societies of global warming exceeding 1.5 °C or 2 °C above pre-industrial levels, greenhouse gas emissions continue to rise. Neither mitigation measures aimed at reducing emissions nor anticipatory actions aimed at preparing territories and adapting our social systems appear robustly deployed. The disconnection between the knowledge and data on climate change and the hesitant and inconsistent action of governments and citizens, as analyzed by Latour (2015), is at the heart of the disorientation and hesitation surrounding climate action today, with the incommensurability of the climate crisis representing one of its greatest challenges. Some authors (Stengers 2015; Bonneuil and Fressoz 2016; Danowski and Viveiros de Castro 2016 or Latour 2017) consider that this disorientation is yet another consequence of the profound transformations caused by the so-called modern model of development, based on the unlimited use of natural resources and indifferent to the planet's environmental limits. The Anthropocene, referred to by these authors as an epoch characterized by a warmer and more populated world with increased risks and hazards, represents a new geological era distinct from the Holocene, where the relationships between humans and the Earth are no longer guaranteed due to the transformations caused by centuries of human activity based on fossil energies. This concept challenges the traditional notion of nature as a static backdrop and instead portrays Earth as an uncertain and threatening reality that can only be understood through the study of its concrete and localized manifestations. We consider this reflection to also apply to climate emergency governance in the context of cities adapting to extreme weather events such as heat waves.

In this sense, in dialogue with theories of resilient adaptation of communities and territories (Meerow et al. 2016; Reeghezza-Zitt and Rufat 2015; Ruíz-Mallén et al. 2022), urban spaces are particularly relevant environments for testing climate-resilient policies. First, because more than half of the world's population lives in cities and, according to the UN-Habitat World Cities Report 2022, this figure is expected to rise to 68% by 2050 (UN-Habitat 2022). Second, cities are spaces particularly vulnerable to heat due to the heat island effect: the rise in temperature caused by the massive presence of materials that absorb temperatures during the day and emit heat at night (Cuerdo-Vilches et al. 2023). Finally, cities serve as diverse and innovative spaces that encourage citizen participation and bring together a heterogeneous mix of individuals, institutions, and knowledge, fostering the potential for collaborative grassroots responses and the creation of shared knowledge.

This chapter focuses on the study of climate shelters, which are extreme heat adaptation devices that are being tested in cities around the world, consisting of the use of pre-existing infrastructures and public equipment, such as libraries, schools, or parks, to accommodate vulnerable people—babies, the elderly, or people with chronic illnesses—during the so-called extreme weather events. While these spaces maintain their previous uses and functions, adapting their infrastructures to the heat by introducing ventilation elements, water points, or vegetation provides thermal, environmental, and social conditions that are welcoming and healthy for people.

Climate shelter networks can be a powerful climate resilience initiative for two reasons. First, they provide heat adaptation for people while optimizing the use of air conditioning—if needed—and can, therefore, be considered a virtuous device from the point of view of climate change mitigation objectives. Second, they are an initiative that can make the climate emergency commensurable. Rather than assuming that climate change is a matter of fact, climate shelters can approach the climate crisis as a reality whose conditions of intelligibility and evidence also depend on community involvement in shared worlds where that evidence is collectively produced and can be meaningful to all actors involved.

The act of transforming an incomprehensible phenomenon into a comprehensible one aligns with Maria Puig de la Bellacasa's perspective on care practices, which emphasizes the importance of giving ethical and political significance to socio-material practices, generating care for overlooked issues, and actively engaging with them to create transformative effects (Puig de la Bellacasa 2017). In this sense, we argue that climate shelters can be both collective infrastructural adaptation equipment to prevent the effects of extreme heat and instances of urban community resilience capable of collectively elucidating the socio-technical terms of such adaptation.

13.2 What Are We Talking About When We Talk About Anticipatory Action and Future Urban Living in the Context of Increasing Temperatures?

13.2.1 Heatwaves in the City: Living the (Future) Climate Change Now

Climate change is causing a general increase in temperatures and an intensification of heatwaves that in urban areas interact with the urban heat island phenomenon, creating a new composite risk characterized by periods of extreme temperatures. Its negative impacts are expected to increase in the future, affecting the habitability of cities and the health of many of their residents, especially the most vulnerable groups (Jones et al. 2015). Indeed, the impacts of exposure to extreme heat are numerous: (i) public health problems such as increased mortality and morbidity, (ii) lack of thermal comfort in indoor and outdoor spaces, (iii) excessive energy consumption and exacerbated greenhouse gas emissions, (iv) infrastructure damage caused by temperature itself or shrinkage and swelling of clay soils, and (v) weakening of biodiversity.

In the context discussed here, anticipatory action is formulated through climatic adaptation policies (Arias et al. 2021).

Those policies are typically implemented at three levels: national, municipal, and building. At the national level, reactive measures such as heatwave alert systems are implemented. Municipal interventions focus on longer-term strategies related to urban planning and design, including green (nature-based), grey (technical infrastructure), and soft (behavioral) solutions. At the building level, improving indoor thermal comfort involves building design, window characteristics, insulation, and occupants' behaviors.

The so-called climate shelters we discuss in this chapter are hybrid devices that consider some of the heat adaptation strategies mentioned above. Although they may involve urban-scale and building-scale transformations, we argue that such strategies have great potential to connect a building's need for thermal comfort with a series of collaborative processes for community and territorial resilience that can be replicated in a wide range of spaces worldwide.

13.3 What Are Climate Shelters, What May They Be, and What is Their Genealogy?

One of the first bibliographic references to climate shelters is the so-called cooling centers (Changnon et al. 1996; Palecki et al. 2001; Widerynski et al. 2017; Nayak et al. 2017; Sehgal and Sehgal 2023). They emerged in the US and are an important resource in extreme heat, especially for people who are homeless or lack adequate cooling in their homes. According to Widerynski et al. (2017), cooling centers are typically air-conditioned or cooled buildings designated as sites for respite and safety during extreme heat events. They can utilize existing infrastructures and staff and be relatively easily implemented. Also, in terms of their management and typology: "they may be a government-owned building such as a library or school, an existing community center, religious center, recreation center, or a private business such as a coffee shop, shopping mall, or movie theatre" (p. 4). In this regard, Black-Ingersoll et al. (2022) distinguish between formal heat shelter buildings or spaces designated by the city council and informal/volunteer air-conditioned shelter spaces open to the public, such as malls, museums, churches, or cinemas not listed by local authorities.

The severe heatwave in Chicago in 1995, resulting in many deaths, was a pivotal moment in shaping heatwave prevention policies in the US, leading to improvements in adaptation and preparedness plans, including the establishment of cooling centers. These measures, along with enhancements to the city's power supply system and the implementation of heat action plans, contributed to a significant reduction in the death rate during the 1999 heatwave compared to 1995 (Klinenberg 1999, 2015; Whitman et al. 1997; Palecki et al. 2001).

The urban heat island is considered an aggravating factor in the Chicago disaster, but climate change is barely mentioned. In the second edition of the book *Heat Wave: A Social Autopsy of Disaster in Chicago*, Klinenberg (2015) added a new preface reflecting how climate change has made extreme weather events in urban centers a global challenge. According to Klinenberg, addressing this challenge requires a commitment to climate-resilient infrastructure changes rather than just relief responses. Additionally, he highlights how social ties, cohesion, and urban dynamism in some working-class neighborhoods in Chicago played a crucial role in saving lives during the 1995 heatwave. That was made possible by a robust social infrastructure of local commerce, public facilities, dense community ties, and organizations that fostered mutual support and intense community resilience.

In this sense, increasingly frequent and intense heatwaves have led local governments worldwide to develop heat adaptation, extreme heat preparedness, and response actions. These actions include, for example, the setup of the so-called cooling centers.

More than 1402 cooling centers in the US have been identified from publicly available sources (Adams et al. 2023). There are also many in Canada and a few in Europe and Australia. Cooling centers are also being built in South Asian countries such as India (Knowlton et al. 2014). However, the existing cooling centers are expensive and depend on a stable electricity supply for the air conditioners to function correctly. Other limitations identified by the literature include the following: (a) some people are not aware of the existence of cooling centers in their city (Alberini et al. 2011); (b) accessibility to cooling centers can be difficult for some citizens: city plans include cooling centers but not transportation to reach them (Bedsworth 2009); (c) some people refuse to go in a cooling center in the event of a heatwave because they do not like the idea of sleeping in a dormitory with nothing to do (Kosatsky et al. 2009; Sampson et al. 2013); and (d) most people do not see themselves vulnerable enough for cooling centers, considering they are not ill (Kosatsky et al. 2009; Sampson et al. 2013).

Despite the need to adapt the city to heat waves, current cooling centers do not meet population needs. Studies show that outdoor cool sites like public pools and parks are used more frequently during heatwaves than indoor cooling centers (Nayak et al. 2017). Public reluctance toward climate shelters may stem from the fact that they are not associated with spaces of well-being, sociability, learning, and community life. Instead, they evoke images of exceptionality and urgency, like flood, fallout, and air raid shelters.

In other regions, noteworthy initiatives, such as those promoted by some local governments in partnership with stakeholders and citizen movements, seek resilient adaptation of small-scale urban services and infrastructures in the context of climate change. One of these initiatives is the Climate Adapted People Shelters (CAPS), a project aimed to reimagine Sydney's bus shelters as Climate Adapted People Shelters through tenets of co-design and co-production that adapt bus shelters to heat (Jacobs et al. 2018).

Nevertheless, since 2019, the notion of climate shelter connects to a specific type of urban initiative for resilient adaptation to heatwaves, explored in Barcelona. This initiative aimed to transform and adapt some facilities, such as public schools, to extreme heat and thus provide thermal comfort spaces—as well as protection, rest, and safety spaces accessible to all—to citizens during extreme weather events (Vilar and Barnada 2020; Baró et al. 2021; Satorras-Grau et al. 2023).

While cooling centers serve as emergency shelters for vulnerable citizens during extreme weather events, the climate shelters in Barcelona are everyday living spaces that adapt their infrastructures and communities to a new world threatened by extreme heat.

13.4 Barcelona Climate Shelters Network

13.4.1 Characteristics and Limitations

The Barcelona Climate Shelter Network initiative launched by the Barcelona City Council in 2019 began to be deployed in the summer of 2020. It consists of a set of pre-existing municipal facilities and public spaces—such as collective social centers, municipal sports facilities, museums, public libraries, schools, parks, and gardens—which, while maintaining their regular uses or functionalities, can be activated, and used otherwise as places to shelter from extreme heat. Barcelona climate shelters can be indoors or outdoors; the former does not necessarily have to be air-conditioned, but they must guarantee a setpoint temperature of 27 °C in summer. The latter should provide greenery, shade, benches for seating, and fountains.

Barcelona is a Mediterranean city with hot and humid weather in summer and is particularly exposed to the heat island effect (Martín-Vide and Moreno-García

2020). Some research indicates that, in the period 1990–2004, mortality in Barcelona soared on extreme heat days by up to 27% (Tobías et al. 2014). Indeed, according to data from the Public Health Agency of Barcelona—ASPB, according to its initials in Catalan—in the 1992–2015 period, extreme heat caused more than 3500 deaths. When temperatures were high, the mortality risk increased by 14% in women and 4% in men (Marí-Dell'Olmo et al. 2022). More recent research indicates that mortality risk in Barcelona on the hottest days also varies unevenly within the city according to age, gender, social class, and neighborhood biases (Ingole et al. 2020). Moreover, climate predictions foresee increasingly intense, prolonged, and recurrent heatwaves in Barcelona (Altava-Ortiz et al. 2017; Pereira et al. 2017; Arellano Ramos et al. 2022).

However, although climate shelters were initially conceived by the City Council as an urban adaptation measure to heatwaves, since 2023, the network has also been attentive to cold waves to protect vulnerable citizens from low temperatures during the winter season. This innovation has not been without controversy, and one of the informants we interviewed for this research, a technician involved in the deployment of the network, shared with us his doubts about it:

> I am very critical of winter shelters; I think they have a radically different logic. Winter season is a different world; those who can die in winter in a city are the people who are homeless. And in this sense, I think there is a gap to be bridged between the reality of winter and the devices that deal with that reality. Summer and winter are different realities and cannot be treated as equivalent (Miquel, interviewed 1).

In this sense, even though initially the climate shelters were only activated in heat-related episodes—there was an activation protocol that operated during the so-called preventive phase of heat waves—today the City Council considers that the indoor spaces are permanently activated—all year round—while the outdoor spaces are activated only during the heat prevention phase, from June 15 to September 15.

According to the Climate Change and Sustainability Office of the Barcelona City Council website, a climate shelter must fulfill the following characteristics:

(a) The indoor spaces—climate-controlled or not—must provide a setpoint temperature of 27 °C in the summer and 19 °C in the winter.
(b) They are designed for people vulnerable to the heat and the cold (children, elderly people over 75 years old, and people with chronic illnesses).
(c) They are not designed for people who need medical care.
(d) They must be easily accessible and safe, providing comfortable rest areas (chairs and benches) and free water.
(e) The urban parks, which form part of this network, have abundant urban greenery (surface area > 0.5 ha) and an NDVI index (which indicates the quality of the vegetation) > 0.4. They are accessible to people with reduced mobility and have fountains and seats.
(f) All climate shelter spaces are adequately signposted.

The Barcelona Climate Shelter Network is one of the responses envisaged in the *Climate Plan* (Ajuntament de Barcelona 2018), the strategic document with which the Barcelona City Council establishes a roadmap for its climate change adaptation and mitigation policies for 2018–2030. This document focuses on four strategic axes. First, mitigation policies are understood as the set of actions aimed at reducing greenhouse gas emissions. Second is resilient adaptation, i.e., actions aimed at preparing the city for the effects of climate change, reducing the vulnerability of citizens and urban infrastructure. Third, climate justice places the most vulnerable groups at the heart of climate action. Fourth is promoting citizen involvement in climate action and governance by committing to democratic and participatory climate policy governance. Indeed, action line 3 of the *Climate Plan* is dedicated to preventing excessive heat. It proposes "improving thermal comfort in the city's public spaces and buildings and creating a network of places to take shelter in high temperatures, to protect people's health and provide special care for the most vulnerable groups" (p. 72).

Through this initiative, the Barcelona City Council aimed to provide all city residents with a climate shelter space within 10 min of walking distance from their homes. Indeed, according to data presented by Barcelona City Council at a press conference on 30 June 2023, 97% of Barcelona's citizens have a climate shelter less than 10 min walk from their home, while 58% have one even closer, just five minute walk from their home (see Fig. 13.1).

Methodologically, the network was designed firstly through benchmarking work inspired by similar projects already in existence in Paris—the *Paris Climate Action Plan* (City of Paris, 2018) set a goal that by 2020, all Parisians could be within a

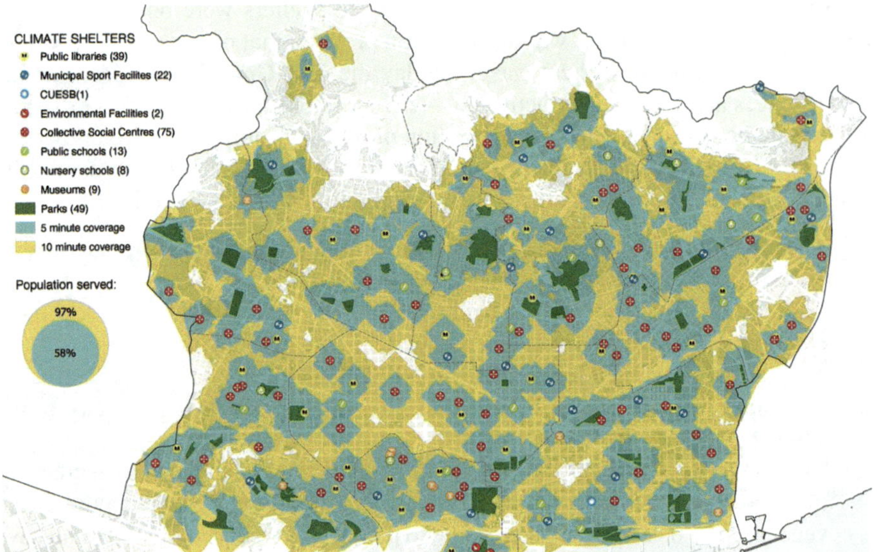

Fig. 13.1 Climate shelter map in Barcelona for the summer of 2023. *Source* Barcelona Regional Agency

seven-minute walk of a cool island (here understood as a "fresh" outdoor space)—and in cities in the US and Australia, and secondly through fieldwork carried out to identify and map potential shelter spaces.

The network has accelerated pace quickly, during which the COVID-19 pandemic also interfered with its development. In 2020, the network was deployed for the first time in the city, making 70 climate shelters available to the public. In 2021, the network reached 155. Moreover, in the summer of 2023, Barcelona expanded its network of climate shelters to more than 200 spaces: 227 climate shelters for the summer and 141 for the winter.

However, in the words of another informant—also a technician involved in the network design—the municipality's obstinacy with the quantitative growth of the network has its ups and downs:

> The local authorities maintain a discourse on the proliferation of the network; however, in the summer of 2021, the City Council carried out a survey asking the level of public awareness of the network. The results showed that 20% of citizens were informed about the network, but only 8% had used it. Therefore, I believe that the City Council should emphasize on raising awareness of the network and once it has become a recognized resource for citizens, focus on expanding it (Aina, Interviewed 2).

According to the fieldwork we have carried out—direct observations of some shelter spaces and interviews with actors involved in the project—some of the principles that inspire the Climate Plan are only half fulfilled in this initiative. Below, to illustrate these shortcomings, we share some testimonies from key informants obtained in our fieldwork:

> A climate shelter should be where a person arrives and can stay in a cool environment where someone offers them a glass of water and asks them how they are feeling. But they are not working like that. Barcelona has 210 shelters right now, but when someone asks about the annual budget allocated to these spaces, the City Council has two answers. Either they answer that there is no specific budget or that the budget is that of the city's facilities, parks, and gardens (Josep, Interviewed 3).

> In the beginning, we were very inspired by some experiences tried in Paris, such as the opening of some parks and gardens until midnight or even allnight during heat waves. What happens in Barcelona is that logistical problems arise quickly: either there is no staff, or there is no staff at that time, so some actions that may be necessary are not carried out (...). Unlike, for example, the American cooling centers, Barcelona's network of climate shelters is made up exclusively of public infrastructures. Our spaces already have a timetable and a staff and, ultimately, what often changes in these spaces is that they gain an identifier and a service of which their staff often have no clear idea, apart from knowing that they have to welcome the people who come here to shelter from the heat. However, our research shows that most people who go to public libraries to take shelter from the heat do so because they are refreshing and welcoming spaces, not because citizens know they are climate shelters. The network has been growing, and the City Council has a discourse of increasing the number of shelters, which, in my opinion, can distort the initiative. What happens? You build up the network, but many of these spaces close or limit their opening hours when August arrives. There should be a network that offers territorial coverage, but perhaps instead of being obsessed with creating many, it is more important to cover densely populated areas of the city. The municipality is sometimes hostage to its own words. There has been a certain fascination with the idea of a climate shelter five minutes away from everyone's home, but have you

asked yourself about the conditions of such shelters? Some spaces identified as shelters are halls of sports centers, sometimes hardly welcoming (Miquel, Interviewed 1).

Public libraries have always been a natural climatic refuge. During heat waves, we have always received users looking for a cool place to connect to the internet, consult magazines, and be comfortable. But, at the same time, adapting our spaces as shelters has some deficiencies. According to the network principles, a climate shelter should have a water point that allows users to hydrate themselves. In many of our libraries, the water point is the toilet tap; we do not have proper fountains; only the new libraries have a proper water point (Lara, Interviewed 4).

The testimonies gathered in these interviews highlight the logistical and budgetary weaknesses of the network, which, while showing its value, highlight the complexity that managing such a network can present in the long term. However, some of the challenges identified have been addressed otherwise through a European project on school-based climate shelters recently launched in Barcelona. We discuss some aspects of this project in the next section of the chapter.

13.4.2 Focus on a Participatory Project: The "Blue, Green, and Grey: Adapting Schools to Climate Change Project"

In 2018, the City of Barcelona submitted the project "Blue, Green & Grey: Adapting Schools to Climate Change" to the third call of Urban Innovative Actions (UIA)—an initiative of the European Commission that provides support for innovative measures in the field of sustainable urban development to test new and unproven solutions to address urban challenges. The proposal was selected, and the project was funded with €5 million (€4 M with European ERDF funds and €1 M with funds from Barcelona City Council) to be implemented over three years. Barcelona's schools house more than 120,000 children considered vulnerable, and their buildings and schoolyards have a low level of heat adaptation. According to the technical document "Good practices: climate shelter projects" (Gerència d'Àrea d'Ecologia Urbana & UIA 2022), Barcelona's proposal was presented as an opportunity to improve the thermal comfort conditions of the city's schools and transform them into healthier spaces for students and teachers, taking advantage of the potential of this transformation to bring benefits to all citizens throughout the year, outside school hours, during and outside the school calendar. The project started in 2019 and finished in 2022.

As a pilot test, the project set out to transform 11[1] public schools to adapt them to heat through three types of climate-resilient measures: (i) Green measures, naturalizing the schoolyards incorporating trees and plants; (ii) Blue measures through the provision of water points, fountains and unique places for children to play with

[1] The schools were chosen from a list of all the city schools that applied to participate in the project. For this purpose, different parameters were analyzed: from climatic vulnerability and the presence of vegetation in the urban environment, or the energy efficiency of the buildings (e.g., number of classrooms exposed to the sun), to the type of paving of schoolyards, and others.

water; and (iii) Grey interventions: i.e., architectural solutions on the buildings such as thermal insulation, shading, protection for façades, cross ventilation, and energy generation.

Although the project was led by Barcelona City Council—under the coordination of the Urban Ecology Area—it also included the following partners: Public Health Agency of Barcelona, Barcelona Consortium of Education, Barcelona Cycle of Water—Public Service Provider—Barcelona Institute for Global Health (ISGlobal), Institute for Environmental Science and Technology (ICTA-UAB) and Vila Olimpica School—the only school that participates in the project since its initial design.

The project turned 11 public schools into a prototype space to implement Barcelona's climate shelter initiative in a profound and committed way. Indeed, the main difference between the adaptation of schools and other climate shelter spaces in the city—such as libraries or museums—is that, in the case of schools, there is a relevant physical transformation process and, at the same time, a participatory and pedagogical process that allows us to imagine and enact a community concerned by the climate crisis.

13.4.3 Co-design and Collective Learning

The transformation of the 11 public schools into so-called climate shelters was done through co-design. However, it should be noted that although the project transforms the school buildings and the schoolyards, only the schoolyards are integrated into Barcelona's network of climate shelters. Only schoolyards are considered spaces open to all citizens during heat waves.

A participatory process made it possible to reach a consensus on the design of the new spaces between the educational community—students, families, school staff, and members of neighborhood associations concerned with the school—and the experts and technicians mobilized by the city council.

First, each school chose a steering group that led the project in the school itself, working on the concept of climate change and imagining possible actions in the school related to heat adaptation. Indeed, in Can Fabra School—one of the schools we visited as part of our fieldwork—the steering group diagnosed on a map the school's (see Fig. 13.2) weaknesses in terms of thermal comfort and proposed ideas on how to solve them.

Four participatory sessions were held over two school years. The specific objectives of these participation sessions were threefold: (a) to involve the educational community in the project, (b) to collectively define the measures to be implemented, and (c) to collectively evaluate the impact of the measures carried out. The first two sessions were held in the 2019–2020 academic year. In the first session (September 2019), they discussed and prioritized the challenges that each school had identified in terms of heat adaptation based on a catalog of available solutions drawn up by the City Council. In the second session (January 2020), the educational community helped the team of architects assigned to each school specify the chosen measures

Fig. 13.2 Details of the adaptation process of Can Fabra School. *Source* photo by the authors

and draw up the executive project for the works. In April 2020, the outbreak of the COVID-19 pandemic made it necessary to hold an extra session—held virtually—to finalize some design issues. In October 2020, the third session was held virtually due to the pandemic. In this case, the works were already executed, and the initial evaluation of the measures implemented was discussed. In June 2021, the fourth session (also virtual) was held to evaluate and verify whether the implemented actions were reducing the temperature of the schools. The total number of participants was 1.264 people.

According to Vilar and Barnada (2020), at the end of the participatory sessions, the chosen adaptations were the following: (i) at the building level, to favor cross ventilation; (ii) at the schoolyard level, to incorporate vegetation, water points, and surface treatment—less reflective and more permeable materials and colors. However, above all, what stood out were the mixed solutions that incorporated elements of shade, greenery, and water in the same layout (i.e., pergola with greenery and water diffusers). It should be noted that while most cooling centers in the world operate through air conditioning, the schools converted into climate shelters in Barcelona operate exclusively through passive measures. However, this decision was not free of doubts and questions. According to one of our informants, passive measures would not be sufficient to cool the school buildings:

> Through passive modes, you can go down by 4-6° C, we have sensorised and checked it, but if you are at 33-34° C, you can reach 27-28° C, with the humidity in Barcelona, maybe the thermal sensation is 32° C (Josep, interviewed 3).

In parallel to the participatory process, the adaptation of the schools as climate shelters carried out a pedagogical process aimed at awareness-raising among the educational community in the face of the climate crisis. In the same way that the participation process mentioned above involves the educational community in the school's adaptation to heat, the pedagogical process aims to involve the same educational community in the fight against climate change, involving in climate action a type of actors—for example, children, but also their families—who do not usually inhabit the forums where these issues are discussed.

During the same two school years (2019–2020, 2020–2021), three training seminars were held in the 11 schools for teachers, as well as educational support for 5th-grade students—the educational level leading the project in each school—on the issue of climate emergencies. This training took the form of a teaching guide entitled "The Challenge of the Climate Emergency: An Information Guide and Proposal for Educational Activities"—a tool for teachers to help them communicate the challenge of climate change—a so-called pedagogical suitcase—and a series of classroom activities and workshops with the project's scientific partners (ISGlobal, ICTA-UAB, ASPB). The pedagogical suitcase to face the climate emergency is aimed at students between 3 and 12. It includes materials and educational resources in physical and digital format on how to deal with the climate emergency in the classroom, accompanied by a teacher's handbook. The activities are organized according to age and with the creation of a learning sequence based on three thematic sections: (a) "Let's talk about climate change," (b) "The climate crisis is already affecting us," and (c) "Let's react to the emergency"—activation dynamics. In the first thematic axis (a), some activities present problems associated with climate change, identifying its origin. In the second block (b), there are activities aimed at discovering in an experimental or sensorial way the consequences of global warming, emphasizing how they will affect our lives. The third block (c) aims to motivate students to act against the climate emergency.

The first seminar for the teachers took place in September 2019. It shared scientific information on climate change and provided teachers with educational resources on climate change.

In September 2020, the second seminar took place to deepen the pedagogical use of the teaching initiatives promoted so far and to socialize the concept of climate refuge among the educational community. In October 2021, it was discussed how to give continuity to the project over time, and the pedagogical suitcase was handed over.

Nevertheless, all this explanation would only be incomplete with some empirical register that would allow us to identify critically the transformations carried out in the schools.

On 22 and 23 May 2023, the authors of this text traveled to Barcelona to visit three schools adapted to the heat to find out the extent of the transformations and interview teachers from each school. The schools visited were Escola Can Fabra—which opened in 2017 on the grounds of Fabra i Coats, a former textile factory that closed in 2005 and converted by Barcelona City Council into a robust arts and culture factory in 2012—(see Fig. 13.2), Escola Cervantes—located in Ciutat Vella, the old district of Barcelona and housed in the Antic Palau Dou, a historic building with baroque and neoclassical architecture—and Escola de les Aigües—located in the district of Baix Guinardó and with an institutional legacy linked to the Spanish army. The first two are part of the climate shelters project financed with European resources, while the last one is part of *Transformem els patis,* a municipal project inherited from the former consisting of transforming schoolyards through participatory methodologies.

According to one of Can Fabra's teachers:

> We already had the idea of transforming the schoolyard, and the Climate Shelter's project fitted perfectly with our aims; we could transform some spaces of the school and at the same time raise awareness of how we can respond to the climate emergency. We love the result, but we really appreciate the participatory process. A technical team accompanied us throughout the process, and the children played a vital role, as did the families. We believe participation has been one of the project's strong points. Before attending the participatory sessions, we had done much work listening to the children, the families, and the technical team that accompanied us, and this work, indeed, allowed us to socialize on the issue of the climate emergency. We modified the schoolyard, replacing the sports court with a Mediterranean garden: a sandy area with trees, shrubs, aromatic plants, etc. We also installed a wooden grandstand —which we call the *agora*— a pergola and a shaded area covered by a sunshade. We installed two multifunctional fountains for cooling, and in the building, we installed solar shading between facades —fixed aluminum solar shades— and a mechanical solar shading system: external screens that are activated mechanically.
>
> We have gained shaded areas, more pleasant, more habitable, and naturalized spaces, and we have also ensured that the summer classroom temperature is not as high as it used to be. This project has been an opportunity to transform the space of our school and the mental framework of the entire educational community; we have experienced how, in the face of a serious problem, such as the climate emergency, there can be a real response that involves a transformation of spaces and that also has a direct consequence in increasing our well-being. The physical transformation needed to make all the participatory and pedagogical work visible was very important, and, in this sense, I would like to highlight the relevance of the physical transformations (Anna, Interviewed 5).

The headmaster of Cervantes School (see Fig. 13.3) also insisted on the positive assessment of the participatory process while showing us some of the most successful infrastructural adaptations of his school:

> We are in a relatively disadvantaged area, a very dense neighborhood with very few green areas, and in this sense, we fit in with the criteria of territorial vulnerability that the project specified.
>
> Our schoolyard used to be a basketball court and has been transformed into a greener space with a multifunctional fountain. We have gained in cross-ventilation in the building thanks to straightforward interventions such as making casement windows and installing solar panels on the roof. Both the participatory and pedagogical processes have been precious; the technicians and mediators sent by the municipality helped us to ground the children's wishes and, in this way, we went from an initial imaginary that immediately thought of a swimming pool to a very subtle type of action that I believe has improved the habitability of the school

Fig. 13.3 Blue, green, and grey solutions at Cervantes School. *Source* photo by the authors

> during heat waves. However, I must also point out that discussing the climate with the children triggered ecological anxiety among many. It, therefore, remains an open question that we must continue to work on (Marta, Interviewed 6).

In short, the main difference between most of the spaces that compose the network of climate shelters in Barcelona and the schools adapted to climate change through the European project is that, in the latter, the issue of climate emergency can be elaborated collectively as shared evidence. We will call this modality of co-construction of climate shelters participatory climate shelters.

13.5 The Notion of Participatory Climate Shelters Regarding Anticipatory Action and Urban Futures

13.5.1 Beyond the Limitations, to What Extent is Barcelona's Network of Climate Shelters an Example of Urban and Community Resilience?

In a context where scientific evidence is not strong enough to guide a collective response to climate threats and conspiracy theories about global warming abound, is it sufficient to adapt urban infrastructures without providing spaces for collective rebuilding in the new climatic regime? According to the anonymous text attributed to the Invisible Committee (2009), the evidence goes beyond logic and reasoning; it resides "on the side of the sensible, on the side of the worlds" and is what is shared (p. 31). How can climate shelters counter the crisis of sensibility that hinders collective responses to the climate crisis?

We use the concept of the "crisis of sensibility," like Morizot (2022), who describes it as both a crisis of our productive relations with living environments and a crisis of collective and existential relations (p. 17). Morizot highlights a Kantian sensibility that separates humans from non-humans, treating the latter as subjective projections of the former. This concept helps us understand the rupture of ties between humans and between humans and territories. To address this, we must understand territories otherwise, no longer as transcendent scenarios in which human life takes place but as a living world with which we are bound by deep relations of interdependence and with which we are obliged to compose our ways of life for habitability.

Morizot's idea connects with the work of the late Latour (2022), particularly with his appeal to the vital role that affect plays in responding to the climate crisis. In Latour's words, "If you do not have the affects capable of metabolizing the ecological situation, everything becomes too hard (…) and the work is overwhelming" (p. 98).

Therefore, in our view, heat adaptation policies require both an infrastructural shift and a new sensibility that allows us citizens to reconnect with other human and non-human entities with whom we maintain relationships of interdependence—often imperceptible—but fundamental for the sustainability of our lives. Moreover, for this, we understand that we need spaces of collective experimentation where climate threats are made sensible in two necessarily complementary ways: (i) as measurable threats that we—an alliance of academic, professional, and social expertise—can recognize and identify together; and (ii) as a new climate regime that oblige us to act in our daily living environments to adapt our territories to the consequences of climate change and thus limit its impact.

In our view, some of the challenges we have outlined have already been tested in Barcelona through the mentioned project that adapted 11 public schools as climate shelters.

13.5.2 An Anticipatory Community?

Section 13.2.1 of this chapter discusses how climate shelters can serve as an urban anticipatory device in response to climate change. Two issues are highlighted. On the one hand, Barcelona City Council's Network of Climate Shelters addresses the climate crisis from the present, when many citizens and organizations still imagine it as a distant issue: it would indeed be a probable future, but above all, a future. On the other hand, the network itself, especially in its school composition, positively contributes to build a more resilient citizen community that can be considered as an anticipatory community. As explained in González Arellano's chapter in this book, an anticipatory community is a group of actors who share a common anticipatory practice: they combine a shared perspective on the future with a specific enactment of that perspective. But what kind of anticipatory community resonates with Barcelona's network of climate shelters? González Arellano (2025) proposes five types of anticipatory communities: (i) repairers—those who respond quasi-random phenomena by designing and implementing protocols to deal with events with known patterns of behavior; (ii) managers—a community that seeks to introduce modest changes that will improve the functioning of the system in the face of events based on regular patterns; (iii) anticipators-strategists—actors with a very open perspective on predictive models and interested in the future scenario building; (iv) visionary-advisors—collectives that share a techno-scientific rhetoric to construct medium and long-term scenarios; and (v) utopian-prophets—those who share concerns about the long-term future, with positions between utopian and collapse theories. Of course, these typologies are ideal models and doesn't exist in a purely discrete urban form. But, at least, in our view, each typology has certainly played a role in the historicity of the Barcelona City Council's Climate Shelter Network. For example, initial Barcelona's school individual-based initiatives to transform schoolyards and adapt them to the heat and sunshine could be thought of as a community of repairers dealing with local increase of temperature due to climate change and urbanization. The phase in which eleven public schools in Barcelona participate in the European project "climate shelters in schools," adapting to the heat through subtle actions aimed at improving thermal comfort can be thought of as the so-called community of managers. Moreover, the technicians and politicians who, based on different climate scenarios, have opted to build a network of climate shelters in Barcelona could be thought of as visionary-advisors. Specially since this type of public policy was completely new in Spain. Finally, On the other hand, the current spokespersons of the Barcelona's network of climate shelters who meet on a regular basis to discuss and imagine the terms of the Barcelona's public facilities adaptation to climate change can be considered anticipators-strategists as they have a good understanding and vision of the network complexity and are able to identify relevant interactions between components, redundancies and institutional or social weaknesses which make it difficult to implement, consolidate and develop such a public policy.

13.6 Final Remarks: Climate Shelters Are Open to All People

On July 10, 2023, as we were writing these lines, the world's media echoed a news item published by the World Meteorological Organization (WMO), which reported that, according to preliminary data, during the first days of July, the world had experienced the hottest week on record. In the words of Christopher Hewitt, Director of Climate Services at WMO, "We are in uncharted territory, and more records can be expected to fall (...) this is worrying news for the planet" (WMO 2023).

As we argued in the previous section, apart from the specific threats that extreme heat poses to the health and well-being of the most vulnerable groups—as well as to the thermal comfort of the population as a whole—we consider that a significant obstacle in the field of anticipatory actions regarding heatwaves has to do with a certain collective inability to make climate change concern us and not just worry us.

Although most institutions, as well as some authors, consider climate shelters as a service aimed at hosting vulnerable populations during specific meteorological events—such as heat waves—based on our research focused on the case of Barcelona, we believe that climate shelters can be much more powerful and decisive spaces both in the fight against climate change and in the adaptation of cities to heat.

We argue that climate shelters have the virtue of operating on the logic of anticipatory action by assuming that climate change is occurring in the present and not just in a still uncertain future that we can continue to wait and not act upon. In this sense, the anticipatory and almost visionary character of climate shelters allows us to think of them as future infrastructures operating in the present.

Nevertheless, why do we consider that a public network of climate shelters such as Barcelona's is relevant to deal with such a sensitive issue as adapting cities to heat?

On the one hand, in a context of uncertainty, where the climate crisis may be driven inequitably, through private solutions that leave critical sectors of the citizenry behind, a public network may bring collective, democratic, and inclusive solutions that have the potential to strengthen the resilient adaptation of communities and infrastructures about the climate crisis.

On the other hand, climate shelters, as a scalable adaptation device, pool energy resources in a crisis that calls for sobriety, makes climate action commensurable, and can involve the population in issues traditionally restricted to the field of experts. However, it is also a device that is easily replicable and adaptable to other types of territories—even non-urban—and latitudes—dealing with the cold or the heat.

Considering the European project that made it possible to adapt 11 schools in Barcelona through participatory methods involving heterogeneous actors from the educational community and the neighborhood, we argue that spaces linked to cultural and educational practices are particularly suitable environments for adaptation as climatic shelters. Thus, despite the difficulties and budgetary or logistical obstacles detected (highlighting problems with maintenance or the difficulty of arbitrating practices with radically heterogeneous opening hours in the same space, which can often strain the calendars and forms of governance of the facilities), we consider climate shelters to be a powerful initiative to anticipate the future. In this sense, we want to conclude this chapter by asking a speculative question that is also a wish: Will climate shelters be able to become a public infrastructure, analogous to public transport, capable of accommodating most citizens during heat waves regardless of their vulnerability?

References

Adams QH et al (2023) Examining the optimal placement of cooling centers to serve populations at high risk of extreme heat exposure in 81 US cities. Public Health Rep 138(6):955–962. https://doi.org/10.1177/00333549221148174

Ajuntament de Barcelona (2018) Pla Clima 2018–2030. https://www.barcelona.cat/barcelona-pel clima/sites/default/files/pla_clima_cat_maig_ok.pdf

Alberini A, Gans W, Alhassan M (2011) Individual and public-program adaptation: coping with heat waves in five cities in Canada. Int J Environ Res Public Health 8(12):4679–4701. MDPI AG. Retrieved from https://doi.org/10.3390/ijerph8124679

Altava-Ortiz V, Barrera-Escoda AA, Amaro J (2017) Escenaris climàtics regionalitzats a l'Àrea Metropolitana de Barcelona (Projecte ESAMB). Projeccions estadístiques regionalitzades, Fase II. Servei Meteorològic de Catalunya & Departament de Territori i Sostenibilitat, Generalitat de Catalunya

Arellano Ramos B, Roca Cladera J, Serra de Larrocha C, Martínez Santafé MD, Biere Arenas RM, Lana Pons FJ (2022) Olas de Calor en la ciudad de Barcelona: 1971–2020. ACE: Archit City Environ 17(50):1–24. https://doi.org/10.5821/ace.17.50.11684

Arias P, Bellouin N, Coppola E, Jones R, Krinner G, Marotzke J, Zickfeld K (2021) Climate change 2021: the physical science basis. Contribution of working group I to the Sixth Assessment Report of the Intergovernmental Panel on Climate Change; technical summary

Baró F, Camacho DA, Del Pulgar CP, Triguero-Mas M, Anguelovski I (2021) School greening: right or privilege? Examining urban nature within and around primary schools through an equity lens. Landsc Urban Plan 208:104019. https://doi.org/10.1016/j.landurbplan.2020.104019

Bedsworth L (2009) Preparing for climate change: a perspective from local public health officers in California. Environ Health Perspect 117(4):617–623. https://doi.org/10.1289/ehp.0800114

Black-Ingersoll F, De Lange J, Heidari L, Negassa A, Botana P, Fabian MP, Scammell MK (2022) A literature review of cooling center, misting station, cool pavement, and cool roof intervention evaluations. Atmosphere 13(7):1103. MDPI AG. Retrieved from https://doi.org/10.3390/atmos13071103

Bonneuil C, Fressoz JB (2016) The shock of the Anthropocene: the earth, history and Us. Verso Books

Changnon SA, Kunkel KE, Reinke BC (1996) Impacts and responses to the 1995 heat wave: a call to action. Bull Am Meteor Soc 77(7):1497–1506. https://doi.org/10.1175/1520-0477(1996)077%3c1497:IARTTH%3e2.0.CO;2

Cuerdo-Vilches T, Díaz J, López-Bueno JA, Luna MY, Navas MA, Mirón IJ, Linares C (2023) Impact of urban heat islands on morbidity and mortality in heat waves: observational time series analysis of Spain's five cities. Sci Total Environ 890:16441. https://doi.org/10.1016/j.scitotenv.2023.164412

Danowski D, Viveiros de Castro E (2016) The ends of the world. Polity Press

Gerència d'Àrea d'Ecologia Urbana & Urban Innovative Actions (2022) Bona pràctica: projecte refugis climàtics. https://bcnroc.ajuntament.barcelona.cat/jspui/handle/11703/127541

González Orellano S (2025) Cities as anticipatory systems. In: González Orellano S, Gandlgruber B (eds) Cities as anticipatory systems. Springer

Ingole V, Marí-Dell'Olmo M, Deluca A, Quijal M, Borrell C, Rodríguez-Sanz M, Achebak H, Lauwaet D, Gilabert J, Murage P, Hajat S, Basagaña X, Ballester J (2020) Spatial variability of heat-related mortality in Barcelona from 1992–2015: a case crossover study design. Int J Environ Res Public Health 17(7):2553. https://doi.org/10.3390/ijerph17072553

Jacobs B, Schweitzer J, Wallace L, Dunford S, Barns S (2018) Climate adapted people shelters: a transdisciplinary reimagining of public infrastructure through open, design-led innovation. In: Fam D, Neuhauser L, Gibbs P (eds) Transdisciplinary theory, practice and education. Springer, Cham. https://doi.org/10.1007/978-3-319-93743-4_17

Jones B, O'Neill BC, McDaniel L, McGinnis S, Mearns LO, Tebaldi C (2015) Future population exposure to US heat extremes. Nat Clim Chang 5(7):652–655. https://doi.org/10.1038/nclimate2631
Klinenberg E (1999) Denaturalizing disaster: a social autopsy of the 1995 Chicago heat wave. Theory Soc 28(2):239–295. http://www.jstor.org/stable/3108472
Klinenberg E (2015) Heat wave: a social autopsy of disaster in Chicago. University of Chicago press, Chicago
Knowlton K, Kulkarni SP, Azhar GS, Mavalankar D, Jaiswal A, Connolly M, Ahmedabad Heat and Climate Study Group (2014) Development and implementation of South Asia's first heat-health action plan in Ahmedabad (Gujarat, India). Int J Environ Res Public Health 11(4):3473–3492.https://doi.org/10.3390/ijerph110403473
Kosatsky T, Dufresne J, Richard L, Renouf A, Giannetti N, Bourbeau J, Julien M, Braidy J, Sauvé C (2009) Heat awareness and response among Montreal residents with chronic cardiac and pulmonary disease. Can J Public Health 100(3):237–240. https://doi.org/10.1007/BF03405548
Latour B (2015) Waiting for Gaia: composing the common world through arts and politics. In: Yaneva A, Zaera-Polo A (eds) What is cosmopolitical design? Design, nature and the built environment. Routledge, pp 21–33
Latour B (2017) Facing Gaia: eight lectures on the new climatic regime. Polity
Latour B (2022) Habiter la terre. Entretiens avec Nicolas Truong. Les Liens qui Libèrent et Arte éditions
Marí-Dell'Olmo M, Oliveras L, Arechavala T, Ariza C, Borrell C, Font-Ribera L, Gómez-Gutiérrez A, González-Marín P, Grau E, Mercuriali L, Montalvo T, Olabarría M, Pérez C, Portaña S, Realp E, Rico M, Sánchez-Martínez F, Valero N, Villalbí JR (2022) Canvi climàtic i salut a la ciutat de Barcelona. Agència de Salut Pública de Barcelona
Martín-Vide J, Moreno-Garcia MC (2020) Probability values for the intensity of Barcelona's urban heat island (Spain). Atmos Res 240:104877. https://doi.org/10.1016/j.atmosres.2020.104877
Meerow S, Newell JP, Stults M (2016) Defining urban resilience: a review. Landsc Urban Plan 147:38–49. https://doi.org/10.1016/j.landurbplan.2015.11.011
Morizot B (2022) Ways of being alive. Polity Press
Nayak SG, Lin S, Sheridan SC, Lu Y, Graber N, Primeau M, Rafferty CJ, Hwang SA (2017) Surveying local health departments and county emergency management offices on cooling centers as a heat adaptation resource in New York state. J Community Health 42(1):43–50. https://doi.org/10.1007/s10900-016-0224-4
Palecki MA, Changnon SA, Kunkel KE (2001) The nature and impacts of the July 1999 heat wave in the midwestern United States: learning from the lessons of 1995. Bull Am Meteor Soc 82(7):1353–1368
Pereira SC, Marta-Almeida M, Carvalho AC, Rocha A (2017) Heatwave and cold spell changes in Iberia for a future climate scenario. Int J Climatol 37(15):5192–5205. https://doi.org/10.1002/joc.5158
Puig de la Bellacasa M (2017) Matters of care: speculative ethics in more than human worlds. University of Minnesota Press
Reeghezza-Zitt M, Rufat S (2015) Resilience imperative: uncertainty, risks, and disasters. ISTE Press—Elsevier
Ruiz-Mallén I, Satorras M, March H, Baró F (2022) Community climate resilience and environmental education: opportunities and challenges for transformative learning. Environ Educ Res 28(7):1088–1107. https://doi.org/10.1080/13504622.2022.2070602
Sampson NR, Gronlund CJ, Buxton MA, Catalano L, White-Newsome JL, Conlon KC, O'Neill MS, McCormick S, Parker EA (2013) Staying cool in a changing climate: reaching vulnerable populations during heat events. Global Environ Change Hum Policy Dimensions 23(2):475–484. https://doi.org/10.1016/j.gloenvcha.2012.12.011
Satorras Grau M, Lara-García Á, Ruiz-Mallén I, Del Moral L, Berraquero-Díaz L, Oteros-Rozas E, March H (2023) Social involvement on urban climate action and governance: lessons from Barcelona and Seville. BAGE (96). https://doi.org/10.21138/bage.3241

Sehgal NK, Sehgal AR (2023) Spatial access to cooling centers in the city of Boston. J Clim Change Health 11:100231. https://doi.org/10.1016/j.joclim.2023.100231

Stengers I (2015) In catastrophic times: resisting the coming Barbarism. Open Humanities Press/Meson Press

Tobías A, Armstrong B, Gasparrini A, Diaz J (2014) Effects of high summer temperatures on mortality in 50 Spanish cities. Environ Health 13(1):1–6. https://doi.org/10.1186/1476-069X-13-48

UN-Habitat (2022) Envisaging the future of cities—world cities report 2022. UN-Habitat, Nairobi, Kenya

Vilar M, Barnada J (2020) We are transforming Barcelona's schools to adapt them to the effects of climate change. Barcelona Soc Rev Coneixement Anàlisis Soc 27:135–147. https://ajuntament.barcelona.cat/dretssocials/sites/default/files/revista-ingles/12_exp_vilar_i_barnada_bcn27_ang.pdf

Whitman S, Good G, Donoghue ER, Benbow N, Shou W, Mou S (1997) Mortality in Chicago attributed to the July 1995 heatwave. Am J Public Health 87(9):1515–1518. https://doi.org/10.2105/ajph.87.9.1515

Widerynski S, Schramm P, Conlon K, Noe R, Grossman E, Hawkins M (2017) The use of cooling centers to prevent heat-related illness: summary of evidence and strategies for implementation. Climate and Health Technical Report Series. Climate and Health Program, Centers for Disease Control and Prevention

World Meteorological Organization (2023) Preliminary data shows the hottest week on record. Unprecedented sea surface temperatures and Antarctic sea ice loss (July 10, 2023). World Meteorological Organization. https://public-old.wmo.int/en/media/news/preliminary-data-shows-hottest-week-record-unprecedented-sea-surface-temperatures-and

Chapter 14
Conclusion

Bruno Gandlgruber, Salomón González-Arellano, Nora A. Morales Zaragoza, and Marco Jaso

Abstract Building a research agenda of cities as anticipatory systems starts with the recognition of the complexity of dynamic processes of transformation in contemporary urban settings that reflect problems of local interrelatedness and global embeddedness. The discipline of anticipation as a new approach to future studies introduces a central role for fundamental uncertainty that shows the importance to visualize existential risks and emerging opportunities associated to long-term horizons. By widening the possibility spaces for future scenarios, different pathways for transformation can be visualized and socially constructed. All relevant stakeholders are included with new forms of agency and governance of the future, including anticipatory communities. The cases in the book provide different kinds of insight based on combinations of conceptualization of futures with varying forms of participation and rule-building. Predictive models are constantly adapted depending on capacities to imagine futures and to introduce these images in the processes of transformation. Anticipatory systems are nested in different levels and types of organizations, connecting thereby images, locations and temporalities. Practices of anticipation are institutionalized formally and informally with heterogeneous levels of futures literacy. Cities as anticipatory systems as a new field of interdisciplinary research still face challenges in topics like anthropocentrism or decolonization.

Keywords Cities · Anticipation theory · Urban futures · Futures literacy

B. Gandlgruber (✉) · S. González-Arellano · N. A. Morales Zaragoza · M. Jaso
Universidad Autonoma Metropolitana, Mexico City, Mexico
e-mail: bgandl@cua.uam.mx

S. González-Arellano
e-mail: sgonzalez@cua.uam.mx

N. A. Morales Zaragoza
e-mail: nmorales@cua.uam.mx

M. Jaso
e-mail: mjaso@cua.uam.mx

S. Gonzalez-Arellano and B. Gandlgruber (eds.), *Cities as Anticipatory Systems*, The Urban Book Series, https://doi.org/10.1007/978-3-031-78162-9_14

14.1 Rethinking Cities and Anticipation Systems

In this book, we have explored different perspectives and modes of the "use of the future" in the urban context, with applications to various sectors and subsystems in other parts of the world, particularly in the Global South. All these examples reinforce the idea that the current challenges of urban transformation are phenomena with high levels of systemic complexity in terms of their local interrelatedness and global embeddedness. Extending explicitly the role of the future in these issues creates additional difficulties related to the consideration of fundamental uncertainty of all expressions of the future and the prevalence of growing existential risks, but also significant opportunities to understand the diverse options of transformation and to build socially and environmentally sustainable pathways towards preferable scenarios in the future. That is why we introduced the proposal of conceiving and studying cities as anticipation systems.

The "discipline" of anticipation distinguishes itself from earlier approaches in future studies for several relevant reasons. Firstly, it recognizes the epistemic change in (social) sciences as a shift from theory building that is exclusively history-based toward inquiries that systematically consider the emerging character of representations of the future, embracing uncertainty and change. In the second place, contemporary future studies have shifted from the traditional focus on quantitative forecasts of specific developments and scenarios toward enriched forms of multi-disciplinary foresight and anticipation exercises with diverse methods and perspectives. In the third place, anticipatory systems and their particular characteristics are analyzed "normatively" without implying standardized solutions towards a (non-existing) optimal outcome. In the fourth place, anticipation theory recognizes the need to integrate all stakeholders and their dynamic mental models to build alternative futures. In fifth place, strengthening anticipatory capacities means challenging established forms of governing the future by including all actors in different participatory and interactive exercises beyond the traditional "expert futures."

The cases presented in this book provide several insights to help us understand urban anticipatory systems better. An analysis of these cases oriented on actors, levels, and scales of anticipation, institutionalization, and futures literacy offers a picture of how city anticipatory systems are developed through open combinations of formal and informal conceptualization, participation, and rule-building processes. Indeed, anticipation can be found at every level, in all sectors and structures of cities, in a more or less organized way. In this sense, we should not confuse the inherent attribute of anticipation as a pre-configurated planning mode. The more we understand the details of the anticipation process, the better the cities' anticipatory systems will be able to guide the transition.

As explained in the introduction, the chapters in this book do not follow a closed methodology of anticipation theory but explore different aspects and subsystems of other cities as anticipatory systems. Seeing the city as an anticipation system is a way to understand how diverse representations of the future guide urban transformation. It provides insight into how citizens and institutions could actively use the future, strengthening anticipatory capacities on different levels. Following the general idea of anticipation systems, all cities as complex systems have the inherent attribute of anticipation as the process of generating images of themselves in the future and using these images for decisions in the present. Urban anticipatory systems are not stable or controlled by specific government entities. Unlike other types of more vertical organizations, cities in particular, and urban systems in general, have developed decentralized, open, diverse, and multilevel anticipatory systems that are continuously changing. Some of the main findings of the different chapters regarding the characteristics of urban anticipatory systems relate to anticipatory actors, levels of anticipation, forms of institutionalization, and futures literacy.

In the first place, the chapters show a great diversity of urban actors involved in different mundane situations where patterns of conceptualization from either individual or collective entities are invariably guided toward building specific narratives of the future that create images of themselves in the future in crucial areas like climate change, the quality life of groups of citizens, resource efficiency, environmental risks, or promoting sustainable development. All these intentions lead to an effort to translate future images into present behaviors. All actors consider aspects of an anticipatory model of a city and adopt their predictive model of themselves and the environment, allowing the system to change. Still, that change is governed by its rhythms and peculiarities. We are especially interested in revealing what type of images preceded specific actions considering the two anticipatory dimensions; the first one is the willingness to imagine a future scenario within a particular disposition of arrangements of actions, and the second one is how they integrate these visions to take action, in other words, their agency or transformative capacity to change the system in function to the future vision that they build and share.

Several chapters have provided an alternative approach to analyze and connect the role of multiple actors, their partnerships, narratives, and citizen participation from a perspective that does not necessarily respond to a geographic denomination but is instead identified as an anticipatory community with complex problems and emergent contexts. We consider it essential to use a mixed multi-perspective focus that involves an epistemological one and the inherited tacit knowledge structure that affects certain actors and practices embedded in specific socio-ecological contexts toward a more long-term visioning that corresponds to the anticipatory literacy model suggested.

Some new topics have emerged in this work, revealing insights toward a deeper analysis of how those visions led to actions. This is the case of the "anticipatory communities" concept, which signals a certain group of actors with a specific anticipatory practice within a regime. Furthermore, we could reflect on hope as a "special force" that catalyzes the actors' impulse to channel their energies toward co-creating a desired future. We could identify different types of urban actors in an anticipatory taxonomy. We see "repairers" and "managers" responding with urban infrastructure works for climate adaptation in Recife, Brazil, or who use the future or "Anticipators" who believe in mushrooms as an alternative to food; locate "Visionaries" that pursue an image of a Healthy City by integrating new indicators of quality of the physical and social environment, stress levels, educational and opportunities in India.

logic or model. Interestingly, in these cases, futures have been disaggregated into horizons with different timeframes and establish the groundwork for greater transformative capacity (for example, through establishing goals in transitional policies and programs).

As observed in previous sections, the capacity for transformation has been observed at various levels. At a micro level, emerging organizations from individual enterprises have found niches for experimenting with new social practices, addressing uncertainty from a market logic (as in the case of mushroom restaurants). Meanwhile, at a meso level, implementing new practices has required the involvement of governmental and state actors, who have been crucial in ensuring the necessary levels of investment and corresponding normative and institutional stability (climate refugees, security services, transportation systems, material substitution). In these cases, the ability to respond to unwanted inertia has been valued as an essential legitimacy capital for citizens.

At a subsequent level of systemic complexity, the identification of unwanted futures at a planetary or macro level, the heterogeneity of interpretation systems by actors with dissimilar interests has hindered the construction of consensus and the implementation of effective transformational policies, programs, or actions. Thus, the developing capacity for change has been announced in the laws or an ideology of change without materializing into social practices with a higher level of coverage or acceptance. However, the introduction of incipient practices and routines could be documented.

For any of the levels and cases studied, it would be highly enriching to understand how developing exploration and use capacities associated with forecasting, foresight, or other creative exploration practices mutually reinforces the evolution and maturation of inherent anticipatory capacities in urban systems.

14.6 The Pursuit of a Research Agenda

Examining anticipatory systems across various cities has consistently highlighted the pervasive nature of crafting narratives and visualizations of the future, particularly about the pressing issues or topics of interest concerning urban planning, governance, or integral components of a global system. Authors have drawn upon various future urban concepts, identifying the most pertinent or defining visions, narratives, or scenarios. They have reflected on their implications for current plan implementation, programs, or targeted actions.

Nevertheless, it becomes evident that one of the foremost challenges in comprehending the development of urban anticipatory systems lies in the comprehensive and precise identification of actors, agents, or stakeholders and in understanding their specific modes of operation in response to future-oriented information (such as data, concepts, images, or narratives). Further elucidation is required regarding the mechanisms facilitating the development of anticipatory capabilities within urban actors and institutions, which is an accumulative process grounded in organizational, group, and citizen capacities. Specifically, there is a need for more research to comprehend the interconnectedness between micro, meso, and macro levels. Authors concur on the necessity for further research into the factors impeding the development of anticipatory capabilities. In this line, it would be beneficial to offer a more comprehensive explanation of the concept of "enacting," wherein future visions gain significance through actionable steps, encompassing various forms of institutionalization such as cultural-cognitive, normative, and regulatory.

It is essential to have a deeper understanding of the interplay between fundamental future literacy capabilities inherent in anticipatory systems and traditional approaches in future studies, such as forecasting and foresight. Additionally, exploring connections with other creative tools utilized for future-oriented thinking would enrich our comprehension.

Finally, all the cases we have studied in this book are anticipatory systems of human social systems. This is an anthropocentric approach that, if we want to overcome, we will have to adopt an approach from socio-ecological systems and the actor-network theory to incorporate non-human actants such as other species (animals, plants) in the analysis. Finally, we consider it essential that in times of uncertainty and unprecedented socio-ecological transitions, it is necessary to systematically and explicitly incorporate a justice perspective in the study and pragmatics of anticipatory systems. Finally, the diverse contributions in this book show the importance to understand the role of hegemonic processes of futures building and the possibilities for decolonization of the future from the Global South.

GPSR Compliance

The European Union's (EU) General Product Safety Regulation (GPSR) is a set of rules that requires consumer products to be safe and our obligations to ensure this.

If you have any concerns about our products, you can contact us on ProductSafety@springernature.com

In case Publisher is established outside the EU, the EU authorized representative is:

Springer Nature Customer Service Center GmbH
Europaplatz 3
69115 Heidelberg, Germany

Batch number: 10732090

Printed by Printforce, the Netherlands